LIGHT RAIL DEVELOPERS' HANDBOOK

LEWIS LESLEY

J.ROSS
PUBLISHING

ISBN 978-1-60427-048-8

Printed and bound in the U.S.A. Printed on acid-free paper
10 9 8 7 6 5 4 3 2 1

Library of Congress Cataloging-in-Publication Data

Lesley, Lewis.
Light rail developers' handbook / by Lewis Lesley.
p. cm.
Includes bibliographical references and index.
ISBN 978-1-60427-048-8 (hardcover : alk. paper)
1. Street-railroads. I. Title.
HE4211.L47 2011
388.4'2—dc22 2011015500

Direct all inquiries to J. Ross Publishing, Inc., 5765 N. Andrews Way, Fort Lauderdale, Florida 33309.

Phone: (954) 727-9333
Fax: (561) 892-0700
Web: www.jrosspub.com

Contents

Preface

Textbooks are difficult to write. Very rarely does anyone read a textbook from cover to cover, like a novel. Normally, short sections are studied for a particular piece of information or to understand an important equation. Because there are overlaps between chapters in the knowledge base, some information has been repeated to avoid the need to refer back (or ahead) to gain a better understanding of a particular point. For the reader who treats this book like a novel, the repetition will seem a luxury. I would ask for forbearance from the technical or professional reader who reads only a part at a time but wants a complete picture without having to read through the whole book to understand the context.

Finally, this book is directed toward the English-speaking world of planning, engineering, transportation, and system promotion. I have attempted to address the book to the different dialect groups and the different administrative regimes. I have therefore tried as far as possible to provide both imperial and metric units and to use technical terms with different variants, to accommodate readers in different countries. I hope that the principles will be understood by all, even if the jargon is not immediately recognizable, and apologize to those who at first read find this difficult.

Lewis Lesley

About the Author

Professor Lewis Lesley is currently a transportation consultant. He graduated from King's College, London with a B.Sc. in mathematics and physics and received his Ph.D. in transportation engineering from Strathclyde University. As a chartered professional engineer and registered professional engineer, he does substantial consulting work for major bus operators, local authorities, and rail operators throughout the UK and the world. He has been a Fellow of the Royal Society of Arts and Industry for nearly 15 years, in recognition of his contribution to public transport developments.

Dr. Lesley has published more than 200 technical papers in refereed international journals and is the author or editor of 25 books. He has held posts as Public Transport Officer, Durham County Council; Reader in Transport, Liverpool Polytechnic; Professor of Transport Science, Liverpool JM University; and Technical Director, TRAM Power Ltd. In addition, he has been a Visiting Professor at Budapest Technical University, Delft Technical University, National University of Ireland in Maynooth, Széchenyi István University, Rice University, and Leeuwarden Hogeschool.

1

Introduction

1.1 The Literature on Light Rail

Railways in general have a devoted following that far outweighs the actual use of rail transport or the market share enjoyed. Light rail is a specialist division of this, and most of the present literature on light rail is aimed at the enthusiast market. This is not to devalue these publications, since many enthusiasts have encyclopedic knowledge of light rail matters and often useful insights into how light rail might be improved (e.g., Joyce 1964; Dunbar 1967; Buckley 1975; Garbutt 1989; Powell 1997; Hass-Klau 2004). Identifying and applying this valuable information is, however, a more difficult matter.

There is also a lot of material on the World Wide Web (e.g., APTA, Lightrail, Lightrailnow, LRTA, Transport 2000, Wikipedia). Most of this is ad hoc and difficult for a practitioner to apply with any confidence to the promotion, planning, design, implementation, or operation of a new light rail system. This book is directed at those charged with designing, building, and operating light rail systems and is based on considerable practical experience, research, and discussions with other knowledgeable practitioners.

Of the books written by professionals for professionals, most come at the subject from a supply-side perspective (Coffey and Kuchwalek 1992; Lesley 1991). In the 1920s in North America and the 1930s in Europe, when car ownership was very low, such an approach made sense, since tramways, streetcars, and light rail systems were selling to captive riders. This is particularly true of those volumes which try to show why one mode of public transport or transit should be used in one circumstance but not in another, especially when based on line haul capacity (Tolley and Turton 1995) (Table 1.1). This is usually based on peak traffic forecasts. In particular, the debate

Table 1.1 Capacity of urban transport modes

Mode	*Maximum capacity (people/hour)*	*Average speed (km/hour)*	*Stop distance (km)*
Bus on road	9,000	16	0.2
Bus on busway	20,000	56	0.8
Light rail transit	40,000	26	0.5
Metro	60,000	32	1.5
Car on road	1,000	19	—
Car on freeway	3,000	72	—

After World Bank reports

about bus, light rail, and metro falls into this category and is based on the fallacious assumption that potential passengers consider all these modes to be of equal merit and quality. Market research shows this clearly not to be the case (Lesley 1974).

A supply-side perspective is inappropriate in developed countries, as public transport is a minority carrier of personal transportation, even in major cities like Berlin, London, New York, Paris, Toronto, and Washington, D.C. The promoter of transit improvements that approaches the subject with a supply-side perspective risks missing the target market.

US Transportation Research Record 1221 (Tennyson 1989) gives evidence to the promoters of light rail about the level of patronage that can be attracted. This was the result of analyzing 40 years of transit data in North America. On the other hand, the UK National Audit Office (2004) report on light rail should make salutary reading for all those promoting better public transport, especially light rail, to avoid the pitfalls of overly optimistic patronage forecasts and cost overruns. The present book partly redresses that imbalance by looking at the considerable market research on what is needed to get car commuters onto (light rail) transit through attraction, rather than punitive measures like parking restrictions and pricing, congestion charging, or other methods of real-time road user taxation. It also considers how to deliver light rail systems that can generate enough operating surplus to service a capital debt, and therefore not need subsidizing by grants from the public purse or taxpayers, and be able to depreciate assets for replacement on a financially sustainable basis.

1.2 The Development of Light Rail

The origins of light rail can be traced back to medieval mines in Europe, where beams of wood were laid longitudinally to make the haulage of wagons easier. These "traams" or wooden balks lent their name to tramways. Later iron plates were used on the timber beams to increase life and give a smoother haul, and with the advent of iron rails and steam haulage came railways. The Stockton and Darlington was first, built and opened by George Stephenson in 1825. In the absence of telecommunications, the spread of new ideas in the 19th century was much slower than today.

Although there is a record of a street railway in 1828 in Baltimore (Dunbar 1967), the birth of light rail can be dated to 1832 when Irishman John Stephenson, no relation to George or son Robert Stephenson of the Liverpool Manchester railway (1830), first laid out a street railway in 8th Avenue, New York. This made the movement of passenger carriages easier for the horses, which then provided all urban motive power. As most urban streets at that time were mainly plain dirt, or in rainy weather mud, rails made a considerable difference in terms of the load that a horse could pull: about 1 tonne on unmade roads and more than 8 tonnes on rails.

This was quickly followed by new horse-drawn street railways in other North American cities. The first street railway in Europe was opened by George Francis Train, an American, in August 1860 in Birkenhead, England. He then built three lines in London in 1861. Soon horse-drawn streetcars and tramways were opened all over Europe. These enabled middle-income families to move from squalid city center conditions to more salubrious housing in the suburbs. Trams allowed them to get around. Various wars and equine illnesses showed the operators of horse-drawn transports that another form of motive power was needed, when the shortage of horses meant that services were curtailed. Experiences with steam traction, and a variety of exotic systems like compressed air, gasoline engines, batteries, cable haulage, and even clockwork motors, failed to identify an economic alternative, until Werner von Siemens in 1882 in Berlin and Frank Sprague in 1888 in America. They, respectively, showed how to use electric current to move vehicles and how to do so safely on public highways. Siemens' trailing wire current collection "troller" system was also used in Orange, New Jersey in 1887.

Sprague's trolley pole system was more economic and efficient than Siemens' trailing wire system. The trolley pole was first demonstrated in 1888 in Richmond, Virginia and to this day still serves the role of transferring electric current to moving vehicle for a few systems (e.g., Boston and Toronto), some industrial cranes, and all trolleybus/coach systems.

The use of electric traction significantly reduced operating costs and made streetcars and tramways the first mass means of transport, affordable by even the poorest. In most of Europe and North America, the funding for the new electric streetcar and tramway systems came primarily from private investors, with commercial companies seeking to reward the capital from operating revenue. In some places, this led to accusations of profiteering and the intervention of public authorities, first to regulate fares and then to take systems into public ownership when cash-strapped operators could no longer maintain equipment.

This changed light rail in most places from a commercial service to having to satisfy political agendas, especially at election times, when a fare rise was seen to be a vote loser. Public bodies often used the operating profits from these systems to subsidize municipal spending, although rarely was enough surplus retained for depreciation and asset replacement; for example, in Liverpool during the 1920s, on average some £200,000 per year was taken from the tramway account to subsidize local property taxes.

Similarly in Europe, tramways were also used for social policy by, for example, making slum clearance possible, allowing people to be moved from inner areas to new housing areas in the suburbs. Municipal trams allowed people moved by slum clearance to still get to their work in the city center, often at subsidized fares.

In North America, many streetcar systems remained under private ownership much longer, and indeed in the 1940s many were bought out by the National Transit Corporation, which was funded by oil, tire, and automobile interests. These systems were then run down, closed, and replaced by buses. This antitrust behavior was investigated by the Federal Commerce Commission. Years later, it imposed a derisory $100,000 fine, by which time transit use was in terminal decline. Ironically, the streetcar systems that survived for example in Boston, Philadelphia, Pittsburgh, San Francisco, and Toronto did so due to the innate conservatism of public authorities, although in all cases the systems were mere shadows of their former networks.

When worn-out infrastructure or rolling stock came up for renewal, some public operators, particularly in Britain, France, and North America, replaced these with cheaper and more "flexible" buses. Sometimes the hybrid trolleybus (coach) was used, but more often the choice led to electric traction being abandoned. When oil was cheap and little was known about air pollution or the limit of oil reserves, this seemed a sensible policy, if it were not for one factor—it ignored the wishes of passengers and the aspirations of the market. The closure of tramway and streetcar systems led to a reduction of patronage, typically about 30%, with ex-passengers buying and using motor cars. This pressured highway authorities to build new roads and enlarge existing streets, at costs orders of magnitude greater than renewing worn-out rails or buying new vehicles.

Even in a "command economy" city like Budapest in the 1980s, the performance of slow on-street tramways delivered quicker door-to-door journeys than faster metros (subways) (Figure 1.1), because of the better accessibility of stations and the short waiting times for frequent service. Arguments for such systems to begin to adapt to high (Western European) levels of car ownership led to the adoption before the fall of the Berlin Wall of a

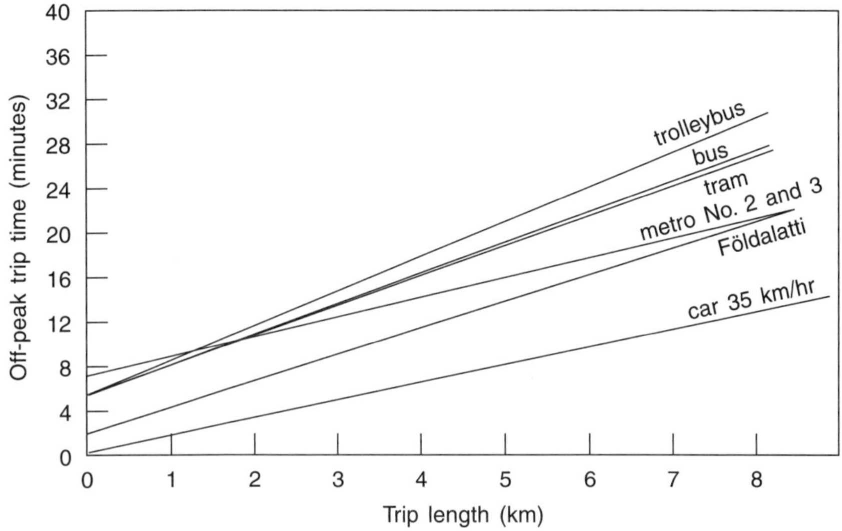

Figure 1.1 Journey time comparisons in Budapest, 1985 (Drawing: L. Lesley)

Gothenburg-style zoned city center, to reduce (through) car traffic, and measures to protect tram tracks from car traffic to maintain operating speeds (Lesley 1986).

By the 1970s, North American cities were realizing the truth of Professor Sir Colin Buchanan's seminal report "Traffic in Towns" (Buchanan 1963) to advise the UK government on future policy for urban transport needs. He showed that the latent demand for private car trips was so high, even in low-density cities like Los Angeles, that there could never be enough road or parking capacity to meet peak travel flows entirely by car. He advocated a twin approach of better public transport and steps to manage car traffic to levels that could be environmentally and socially accommodated. With North American cities reaching gridlock and downtown areas dying for want of acceptable access that bus systems did not provide, municipal authorities saw that rail transit attracted car commuters but could not afford new subways or metros.

Metro or subway systems are expensive. The reasons for this relate to the need to acquire large sites for stations and the cost of tunneling. Pierre Laconte, Secretary General of Union Internationale des Transportes Publiques (UITP), observed that light rail offers 90% of the benefits of a subway at 10% of the capital cost (Laconte 1978). Light rail or modern streetcars were seen as a quick and affordable solution. The image of streetcars in North America bore antiquated connotations and ideas of slow and inefficient operations.

Therefore, the new generic name "light rail" was used to relaunch streetcar transit. By the 1980s, policies of privatization and deregulation were taking the stage politically in many countries. The aim was to remove the patina of public sector inefficiency and pork barrel politics. We will not debate the rights or wrongs of either the public or private sector as the promoter or provider of light rail systems but rather will show how light rail can be made to work through either approach.

The use of open competition is now a cornerstone of the procurement policy for publicly funded projects to improve and reduce the costs of transit services all over the world. In North America, transit remains highly regulated and largely publicly owned. There is also an ongoing debate over the difficulty of introducing innovation in light rail, when risk-averse authorities only seek equipment that is tried and tested.

1.3 Environmental Impacts

Transport consumes resources and impacts the environment in the following ways:

1. Resource extraction to construct infrastructure
2. Resource extraction to provide the motive power
3. Air emissions from operations
4. Noise and vibration
5. Community division

1.3.1 Resource Extraction

Raw materials are mined, refined, and manufactured for use in vehicles, ways, maintenance, and garaging. At the end of their working lives, these must be disposed of. Some materials can be recycled, whereas others are dumped. Both activities have environmental impacts, consume energy, and release emissions into the environment. Fortunately, the size of this resource depletion is small in comparison to the environmental impacts of operating transport systems.

1.3.2 Energy Consumption and Pollution

Transport is an energy-intensive operation and oil dependent. In the developed world, nearly 99% of transport system output depends on oil. Various commentators have focused on looming peak oil, when worldwide production will begin to decline. For the last 10 years, new oil field discoveries have not kept pace with the growth of oil consumption, as more countries seek

"Western" and energy-intensive lifestyles. For transport security, new sources of motive energy are required.

The largest impact of oil consumption, however, comes from the burning of fossil fuels inefficiently. About 75% of the energy contained in oil does not produce movement, but instead is wasted as heat or unburned fuel. Transport is the most energy-inefficient sector of the economy. Electricity generation from fossil fuels only wastes about 60% of input energy. Transport fuels have been cheap enough to be used wastefully, because the price paid reflects the cost of extraction rather than replacement.

Finally, pollution that threatens health and the environment is released by transport systems, primarily road transport. Oil is a hydrocarbon fuel, so burning it in oxygen produces carbon dioxide (CO_2). About 30% of all CO_2 in the developed world comes from transport. The actual amount and the proportion of the total released by transport are growing, as other sectors of the economy adopt energy efficiency measures and renewable energy sources. CO_2 has been linked to global warming and climate change that could fundamentally change the habitat of the world, making the survival of some fauna and flora problematic in many places, especially if sea levels rise.

The waste gases from transport operations include carbon monoxide, oxides of nitrogen, and small particles (PM_{10}). All of these are associated with a number of cardiovascular and lung diseases. Most recently, researchers have linked high levels of PM_{10} with higher risks of women miscarrying.

While there is public concern over deaths and injuries caused by transport crashes, many more people die and are made ill by transport pollution. In the UK, some 3000 people a year are killed in transport crashes. In comparison, 45,000 die from transport pollution diseases.

At a local level, transport operations cause smog, which reacts with sunlight, making the health effects worse. Los Angeles is notorious for this because of a combination of topology and climate. However, "inversion" smog occurs in many cities. In Europe, the EU Commission has issued Air Quality Directives, which set the maximum concentrations of various combustion products. Similarly, the World Health Organization has produced what are considered to be safe levels of urban air pollution, especially where children are exposed. It has been recorded in many cities that children living near busy roads suffer much higher levels of respiratory diseases, including asthma, than children living in the suburbs or in rural areas. Many British cities fail to meet the EU air quality standards.

1.3.3 Noise and Vibration

Road traffic noise is the most widely reported noise nuisance in most cities. The World Health Organization recommends a maximum level of noise of 55 dBA for social living. Many city streets have noise levels of 65 dBA and above, making conversation difficult if not impossible. All vehicles contribute to this noise. For most motor cars, the significant noise generator is now the tires, not the engine, as a result of improved engine and silencer/muffler design. Larger vehicles are noisier, since diesel engines are more than 70 dBA. Better engines and transmissions will not reduce the overall noise level below 70 dBA, since that is the noise generated by rubber tires on the road pavement.

Roads are not perfectly flat, so vehicles bounce as they progress. This bouncing is amplified by the load of the vehicle and nature of the suspension. The bouncing is carried by the road pavement into adjacent buildings, where windows and fittings rattle. There is no record of buildings being structurally damaged by such vibrations, but they are annoying, and occasionally trinkets fall over and break.

1.3.4 Community Division

"The wrong side of the tracks" is a well-known social phenomenon, where different income groups occupy areas separated by railroad tracks in American cities. Highways also can divide communities. As early as 1963, an environmental traffic flow level was defined (Buchanan 1963) for residential areas as under 200 vehicles per hour. Subsequent research shows that how pedestrians cross roads is dependent upon the vehicle flow. The higher the flow, the more likely the crossing will be at right angles. Put the other way around, if pedestrians cross the road at obtuse angles, the traffic is likely to be at an environmentally acceptable level.

Finally, many studies have shown that the number of neighbors people know on the other side of the road is highly correlated with traffic flow. The higher the traffic flow, the smaller the social network, with few people from the other side of the road. Of course, some roads have been built (e.g., freeways and motorways) to exclude pedestrians and to make crossing the road difficult or impossible.

1.3.5 Light Rail Impacts

Light rail systems that offer the same capacity as a busy road with car traffic can operate at longer than one-per-minute intervals, making social cohesion easier to achieve. Because they are electrically operated, no air pollution is emitted in the street, and if powered by renewable generation, they are CO_2 free, need no fossil fuel, and therefore are environmentally and energy sustainable. Electric traction and light rail vehicles are quiet, typically emitting less than 60 dBA, and steel wheels on steel rails produce little in the way of ground-borne vibrations.

1.4 Planning Light Rail

This book will not consider in detail the legal requirements found in different countries, or even cities, for light railways to gain statutory authority to build and operate, since there are so many approaches. Instead, detailed consideration will be given to the processes by which routes can be identified and evaluated, environmentally and economically. These can be applied to most urban areas, irrespective of the statutory requirements. They are based on established demographic, economic, and geographic analyses that allow potential catchment areas to be compared in terms of trip generation and attraction for new light rail lines.

Similarly, the way in which systems can be operated will be evaluated, to provide an acceptable quality, by identifying the optimum use of resources and considering how passengers react to different combinations of service characteristics. These aspects of planning are universally accepted and the principles well documented (e.g., Banister 2002; Bruton 1970; Faludi 2004; O'Flaherty 1997; Pharoah and Apel 1995; Starkie 1976; Tolley and Turton 1995; Vickerman 1991; White 2005). Use is made of generalized costs as a proxy for the complex way in which people compare different travel options and then select the ones to make particular journeys or choose a home location to minimize the travel cost or time for the most important travel—the journey to work.

1.5 Engineering Light Rail

The physical forces created by the movement of (light rail) vehicles, the capacity of ground to carry loads, and the strength of materials and structures are all universal principles of physics and civil engineering. Therefore, the detail of how to determine the best designs and material configurations will be examined, together with different technologies and installations to achieve a safe operating environment. These principles are transferable from one place to another, allowing the light rail engineer to use previous successful experience when considering a new system and therefore reducing the costs and risks of installation. This is the fundamental scientific approach by which humanity has made considerable technological discoveries and inventions.

Safe, reliable, and robust light rail systems that can be afforded should be the objectives for all engineers. This book provides a systematic basis for achieving that and in doing so looks at attempts that were tried but not always successful, thus helping light rail promoters avoid costly mistakes. It also draws on widely available texts on general engineering aspects of highways and railways (e.g., Atkins 1980; Morgan 1973; Profillidis 2000; Salkield 1953; Sharp 1970; Slinn et al. 1998).

1.6 Affordable Light Rail

When light rail systems are being promoted by public authorities, the capital required has to come from a budget provided by taxpayers. This is not limitless. Demands on taxes for public capital works and revenue payments like transit subsidies usually outstrip taxpayers' willingness to fund. Inevitably, a mechanism to justify the use of "free" money, and maybe also operating at subsidized fares, is needed (Lesley 1987). Cost-benefit analysis was developed as one accepted way to demonstrate and justify the nonfinancial benefits of such public investments. These benefits include environmental improvements, reducing congestion, encouraging economic activity, reducing the need to expand the road network, etc.

Politicians also undertake an electoral calculus on which projects will win the most votes in the most places. In this scenario, light rail has a low priority compared to schools, hospitals, and highways. Even in cities with widespread traffic congestion or high levels of auto air pollution, there has not been enough pressure to provide light rail for more than a token diversion of car trips. Rarely is highway capacity removed in such cases. This means that suppressed car trips quickly fill the space vacated by trips diverted to light rail. Therefore, no overall environmental improvement can be measured. There are texts by transport economists that can enlighten this debate (e.g., Glaister 1981; Lesley 1996; Townsend 1969).

In North American light rail systems, capital costs have varied widely between the low achieved in San Diego ($10 million/mile = €5 million/km of first line) (Figure 1.2) and the high for the Buffalo system ($100 million/mile = €43 million/km), with the middle near the European average (€18 million/km).

At the UK Parliamentary Light Rail Inquiry held in 2009, the question of light rail costs was raised by the chairman, Paul Rowan, MP. New light rail construction costs in the UK are about £20 million/km ($50 million/mile) and rising. Andrew Braddock (formerly of Transport for London) reported that the target price of new tramways in France is €14 million/km (approximately $33 million/mile). A recent extension to the Brussels tramway cost €8 million/km ($19 million/mile). Karlsruhe on-street track extensions are €7 to €8 million/km ($16 to $19 million/mile). A senior German tramway director in Stuttgart commented on the French costs: "The French figures often include restructuring/reshaping of the entire street environment. Thus I do not consider them apt for cost comparison of pure permanent way construction. Our figures for a double-track alignment (incl. over-

Figure 1.2 San Diego—an affordable light rail system (Photo: Peter Ehrlich)

head, stops and signaling) vary from €5 to 10 million per km" ($12 to 24 million/mile).

In spite of German construction workers being better paid than British, the German tramway construction cost is less than half those in the UK. This may explain why there are more tramways and extensions being built in Germany than the UK. The new GLUAS system in Galway at €9.5 million/km ($22 million/mile) is consistent with German tramway costs. For light rail to play a larger part in providing an energy-sustainable alternative to urban car trips, construction costs need to be reduced. Ways to do this will be discussed.

On top of this, light rail new systems that do not cover their operating costs need operating subsidies. This assumes that the capital is a free grant and not a loan that has to be repaid. For politicians funding and subsidizing light rail, this can seem expensive, especially when motorists driving cars are an important source of tax revenue. For public finances, therefore, light rail can seem to be a negative investment, needing taxpayers' money to build and operate and reducing the tax take from reduced fuel sales because of trips attracted to light rail transit.

Many US systems have been funded by bond issues underwritten by local sales taxes (increases), typically about 2%. These tax increases are usually approved by a referendum of electors. A study by the University of Texas at Dallas asked electors in Dallas why they had voted to increase sales taxes to fund a new rail project. The answer was that although most electors did not necessarily see themselves as transit users, enough car drivers would be attracted to the light rail system to reduce traffic congestion and make driving easier for the rest (Figure 1.3).

Figure 1.3 Successful first line in Houston (Photo: Mosal Con Hermann)

The advent of low-cost airlines shows that the public sector model for transport is not the only one which can provide an acceptable service for passengers. In places where transit is operated commercially, there is an argument that new light rail systems might be promoted by private companies and financed privately on the basis of generating operating profits from fares. Looking at the economics of light rail from a commercial perspective could help to unlock funding for new projects that otherwise could not be afforded by the public sector. Privately funded light rail systems can also meet municipal aspirations and increase civic pride. To achieve this, there needs to be a genuine partnership, which is usual in other commercial and real estate development.

In this case, the potential profitability of new projects and the dividend that will be paid to investors will determine whether a privately funded system goes ahead. The likely viable route(s) will be through areas that are economically active, with high levels of mobility and congestion, rather than areas needing regeneration, if a project is to be attractive to private investors.

In summary, an affordable light rail system is one that operates at a surplus, can depreciate and replace assets as they wear out, and, should the capital investment be provided by the private sector, services the capital debt. To achieve this, the system and installation should be the minimum commensurate with safe and attractive operations. This means:

- No special structures or tunnels, bridges, or viaducts unless unavoidable
- Capacity to meet the near-term operating needs only
- Alignment and stations to maximize patronage and revenue
- Alignment and traffic priority to minimize journey time and fleet size

Achieving a viable system also requires considerable engineering inputs, as well as accommodations to minimize the negative urban impacts. Such impacts occur while the light rail system is being built and become permanent once it is open. The change of the US President and administration in January 2009, with a more overtly green agenda, could accelerate the construction of new light rail lines in America. This could help reduce dependency on imported oil for urban car trips. In May 2009, an agreement was also reached with the US auto industry to build and sell new cars that get more miles per tank of fuel and create less carbon dioxide and other polluting emissions.

1.7 US Transportation Research Board Light Rail Study

Set against US national policy are a number of anti-light-rail lobbyists, like Wendell Cox, Randal O'Toole, and Bill Vincent (of the Breakthrough Technologies Institute), advocating bus rapid transit as a cheaper option than light rail.

Even serious transport commentators sometimes criticize light rail as being expensive and claim that buses (on busways) could do the same job, for much less money. This was a policy followed in Houston in the 1980s and 1990s with the construction of more than 100 miles (160 km) of high-occupancy vehicle (HOV) roads on six major routes. More importantly, buses are not perceived to be the same by those whose patronage is required to earn the revenue to pay for the operating costs—the passengers who presently drive cars. While ad hoc market assessment studies have been conducted in different cities on the views of residents, and especially car commuters, of alternative forms of transit, none has been as systematic and evidence based as that published by the US Transportation Research Board in 1989 (Tennyson 1989).

Report 1221: "Impact on Transit Patronage of Cessation or Inauguration of Rail Service" covered a 40-year period from 1945. It studied the impact of the closing of (light) rail lines on travel patterns, as streetcar and commuter rail lines were abandoned, to be replaced by buses, which in many cases were also later abandoned. Towards the end of the period, the impact of opening new light rail lines, to replace buses, on travel patterns was also studied. The closure of rail lines, replaced by buses, often with enhanced service frequency, resulted in a loss of transit patronage of between 30 and 40%. The passengers lost to transit transferred to car travel.

In the case of opening new lines, where a comparison could be made with existing bus lines, or where bus lines were replaced by light rail, on a like-for-like basis, rail carried about 40% more passengers than bus lines. The report concluded:

> In most cities served by buses exclusively, transit riding has declined 75 percent over the past 40 years. Exclusive busways have not made much difference absolutely, but they have helped relatively. In 11 areas with updated rail transit facilities, ridership has increased markedly, often by more than 100 percent. In two of these areas, the transit systems are attracting more ridership than they did when gasoline and tires were rationed. It appears that rail transit makes a great difference in ridership attraction, with attendant benefits.

The report continued:

> When these service conditions are equal, it is evident that rail transit is likely to attract from 34 percent to 43 percent more riders than will equivalent bus service. The data do not provide explanations for this phenomenon, but other studies and reports suggest that the clearly identifiable rail route; delineated stops that are often protected; more stable, safer, and more comfortable vehicles; freedom from fumes and excessive noise; and more generous vehicle dimensions may all be factors.
>
> Those engaged in the alternatives analyses (*of transit modes*) and similar studies would be well advised to consider these differential factors before making service recommendations or traffic relief assumptions. Future problems with air pollution, congestion, and funding may all be seriously affected by these considerations.

While there has been no similar comprehensive study in other countries, trams (streetcars) were abandoned in Britain during 1948 to 1962, all replaced by buses. In most cases, the new buses were more comfortable than the 50-year-old trams replaced and ran faster and more frequently. In all cases, the patronage declined immediately by between 30 and 40% and continued to decline at an average of 2% per year and overall by 70% since 1955.

In France, some ten new light rail systems have been opened since 1985. Nearly all of them replaced the busiest city bus lines. On opening the new light rail lines, the patronage jumped within a year by a minimum of 30%, compared to the previous bus services. In Britain, five new light rail lines have opened since 1992. Most of these have been on old or closed urban rail lines. Market research of the patronage of these lines has shown that between 20 and 40% of passengers have transferred from car commuting.

These data indicate that there are qualitative differences between light rail and bus services, which heavily influence customers, and therefore ridership and revenue. In particular, light rail has shown itself able to attract significant traffic from private car commuting. This modal transfer is critical to help cities address and reduce traffic congestion, air pollution, and other environmental problems. In Houston, construction began in June 2008 on five more light rail transit lines, a 30-mile- (48-km-) long light rail system, to supplement the successful first line opened in 2004. HOV lanes meanwhile are being tolled for single-occupant car use, implying the failure of bus lines using them.

1.8 UK National Audit Office and Audit Commission Reports

Recent experience in the UK may also provide some lessons for those around the world promoting new light rail systems. In 2004, the UK National Audit Office, which monitors all spending by the central government, published a report on an investigation into the light rail systems which had been opened in the previous 15 years in Birmingham, Croydon, Manchester, Nottingham, and Sheffield. This report showed that light rail has not generated enough (political) payback for further new projects to be funded. Indeed, those projects were seen as poor value for the money when compared to similar projects in continental European countries, which were about 50% of the capital cost of UK systems that deliver the same transport objectives.

This analysis ignores the way light rail can achieve other objectives (e.g., environmental, economic, traffic, or social). Many of these objectives, however, result in no increase in operating revenue. The UK National Audit Office showed that light rail systems in the UK rarely fully achieved the objectives that were claimed by promoters. Indeed, the use of cost-benefit analysis, which tries to equate these disparate and often subjective elements, rarely shows light rail to be good value for money compared to other transport projects aiming for environmental benefits, like cycleways. Perhaps this is because time savings from light rail systems are often small or negative compared to car trips.

The report also commented on the various and sometimes conflicting objectives the public sector promoters claimed that light rail projects would achieve. Economic regeneration and reducing social exclusion were often cited as key objectives. This is a confusion of cause and effect. The Birmingham light rail transit system passes mostly through the "brownfields" of the deindustrialized "Black Country" area (Figure 1.4). Little economic regeneration has taken place since the light rail line was opened, because of the general state of the economy. This light rail transit system has failed to achieve its projected ridership and revenue targets by about 50%. The operator claims to be losing about £10 million per year but cross-subsidizes from a network of local bus services also operated. This is hardly a good use of transport resources.

As a result of the National Audit Office report, the central government in the UK withdrew grant aid for planned light rail systems in Leeds (Figure 1.5), Liverpool (Figure 1.6), London, and Portsmouth and has forced the promoters of new extensions in Birmingham and Manchester to compete for funding from a Transport Innovation Fund. This is meant to support all capital projects throughout England. Both cities are seeking to use 100% of the yearly allocation, which even if spread over 4 or 5 years would significantly reduce the funding for other transport projects in the rest of England. Scotland and Wales have separate capital funding plans. The vote on a ref-

Figure 1.4 Light rail in Birmingham, UK (Photo: L. Lesley)

Figure 1.5 Aborted Leeds Supertram system (West Yorkshire Passenger Transport Executive artist's impression of Supertram in Castle Square)

erendum in Manchester in December 2008 was 79% against a congestion tax to pay for light rail system extensions costed at £3 billion. Recently, Manchester has been awarded £1.5 billion spread over 5 years for light rail extensions and other projects.

Subsequently, the world financial crisis in the first decade of the 21st century has created a "black hole" in public funding. Finding the money will be difficult if not impossible in the near future for an expansion of light rail,

Figure 1.6 Impression of abandoned Merseytram (Photo: Merseytravel)

as it will mean having to cut schools, police, trash removal, and other publicly provided services. The cash crisis in California is matched by similar crises in Britain and Ireland.

A case study of one failed UK light rail project flags some pitfalls that other promoters would do well to heed. The Audit Commission, which oversees expenditure by local government in England, published a report in January 2008 on the failed "Merseytram" project in Liverpool (Bartlett 2008). This investigation found that some £70 million ($110 million) had been spent on developing the project, additional to the projected £325 million ($500 million) capital cost for a 19-km (12-mile) light rail line. Of this, £28 million was spent on consultants, £17 million on design and management, £15 million on utility diversions, £1.3 million on a (failed) legal challenge against the government, and £0.9 million on rails, which at the time of this writing are stored in Immingham Dock. The promoting authority was the Merseyside Passenger Transport Executive (MPTE = Merseytravel). The chief executive, director general, and chief engineer at the time was Neil Scales.

The MPTE's justification for spending £28 million on consultants' fees was that it did not have the required expertise in-house. This, of course, also means that it did not have the required expertise to supervise the consultants' work, to ensure it was appropriate or value for the money. The utility companies will no doubt be very pleased to have their plants relocated and replaced along the line of the aborted route. Was it prudent to spend a further £1.3 million challenging the government over its decision in November 2005 to withdraw a grant aid offer of £170 million, due to a locally unfundable £155 million gap? The upshot of this is that MPTE borrowed £50 million, has taken £20 million from its reserves, and now has a further annual burden for repayments of more than £5 million, requiring a budget reduction of £22 million over 3 years and a reduction of staff by up to 25% (Audit Commission 2006).

The Audit Commission also criticized the fact that significant sums of cash were committed to the project before Merseytravel had the money to pay for it, and there were currently no "tangible benefits." The commission concluded that:

- Merseytravel did not pay sufficient attention to managing risk and should not have committed resources at the rate it did.
- Merseytravel did not have adequate financial reporting and monitoring arrangements to demonstrate that funds spent on the project represented value for the money.
- Merseytravel did not engage sufficiently with all the councils in the area.
- There was limited opportunity for challenge and too much reliance was placed on too few individuals.

- The expenditure on the project was "not so unreasonable" as to be unlawful.
- Given the various roles of its chief executive, Merseyside Passenger Transport Authority also should have done more to ensure it had access to appropriately independent advice.
- In these circumstances, it remained important that the authority had effective arrangements in place to hold officers—and consultants employed by them—accountable through effective challenge and scrutiny.

The Audit Commission published a Public Interest Report, seen as very serious and reserved only for instances of wrongdoing or impropriety.

The two reports, from the UK National Audit Office and the Audit Commission, have done nothing to bolster public confidence in Britain over the effectiveness of light rail systems or that public authorities can achieve value for money. This may go a long way to explain the coolness of the UK government to fund new projects, despite their popularity with the public, their ability to divert commuter car traffic, and their environmental benefits.

Indeed, a cynic might argue that £70 million could have funded a comprehensive safe cycle network throughout the whole of greater Liverpool and Merseyside, and offered low-cost bikes to those without, to target the 50% of car trips that are less than 5 km (3 miles) in length. Not only would this have reduced traffic congestion and emissions all over the area but it would have led to a less obese and healthier population as a result of exercising more.

2

Characteristics of Light Rail

2.1 Market Perception

There are significant volumes of market research produced over the last 30 years which show that people can differentiate between modes of public transport, rank their qualities, and evaluate their own willingness to change lifestyles to use the various systems (e.g., Butler et al. 1969; McMillan and Assael 1969; Paine et al. 1969; Golob et al. 1972; Hensher 1974; Spear 1976; Stanley 1977; Daly 1981; Dix 1981; Hensher 1989; Pas and Harvey 1997). Usually more than half of the respondents are willing to move and to live closer to an electric rail service for the work commute. About 25% are willing to move to a diesel-powered rail service, while an insignificant number will move their home in order to use a bus service to commute to work, since in many countries the car is the dominant mode of transport to work (Voorhees et al. 1974).

These perceptions arise from a number of different forces. Most European (about 50 cars per 100 population) and North American (about 70 cars per 100 people) families have cars as the first-choice mode of personal transport. The family car provides a standard against which to compare alternatives, in terms of quality of service and cost. Most people would rather sit in their car in a traffic jam than in a bus in a high-occupancy vehicle or bus lane. The "flexible" bus has a consistently low image among car drivers and even captive (carless) riders (Lesley 1974b). There are many reasons for this; some are rationalizations, while others are intrinsic to the performance of bus service or people's perception of service quality.

Although urban bus use has fallen significantly (more than 70% in the US and UK) over the last 50 years as car use has grown, there is one mode of public transport that continues to grow: taxis. In the UK, the highest per capita taxi use is in the poorest quartile of households. The next highest users are in the top income quartile. Taxi fares in real terms have fallen but are still higher than transit fares. Taxis provide a door-to-door service. Among the poorest families, much taxi use is shared, while in the richest, single occupancy is the norm. In the US, there are many towns that only have taxis as public transportation.

There is clearly a role for taxis in serving light rail systems to extend the catchment area, and indeed some transit operators use taxis to supplement or provide night services. At key interchange stations, a taxi stand will also be important for incoming passengers who are strangers to the area.

There also have been significant experiments, some intended and others salutary. Probably the most interesting is the UK new town of Runcorn, where in the 1960s a complete new town was built around a busway (Figure 2.1), with the town development centered around busway stations. The highway/freeway network was built so that journeys in Runcorn would usually be faster by bus than car. It was expected because of this very high-quality bus service that about 50% of the journeys in Runcorn would be by bus. In fact, bus use in Runcorn was no higher (at about 12%) than in other UK towns with a similar population size but no busway (Lesley 1983) and since has continued to fall.

New light rail lines in Europe and North America also have demonstrated that a significant transfer from cars can be achieved quickly when the system has been developed to satisfy market aspirations. Even those lines which were built on a low budget, like the pioneering Skokie Swift (1964) in Greater Chicago, have shown that real market penetration can be achieved when a high-quality service is provided.

Figure 2.1 Runcorn Busway at Shopping City, 1985 (Photo: L. Lesley)

The San Diego Trolley (1981) in California shows both that light rail need not be expensive to construct and that existing rail freight lines can be shared, using time separation to avoid costly track duplication. This makes better use of existing infrastructure. Similarly, the first line of the Manchester Metrolink between Altrincham and Bury, using an old rail right-of-way, quickly achieved more than double the estimate of demand once opened, with about 30% of trips diverting from private cars.

Why the difference between perceptions about bus and rail transit? First, it is worth considering four major characteristics of using public transport: cost, journey speed, frequency, and reliability. Of these, invariably reliability is rated the most important and cost the least. Car commuters pay much more to own and operate a car than the cost of equivalent transit fares. People pay this high cost for the convenience and dependability of car travel, which is important for getting to work.

Rail-based systems are always seen to be permanent, while "flexible" bus systems can be discontinued at short notice, even in regulated systems, as in most North American cities. Moving one's home or changing jobs requires a dependability of transport to get to work. In the UK and US, the car is now the most important mode of transport to work, except in the core of very large cities like London (Table 2.1) and New York. There is also the issue of day-to-day reliability. Getting to work on time is critical for most people, and the experience with bus systems for many people is that they have a lower level of operational reliability than rail. This excludes journey speed and ride comfort.

In the US, where public transportation (transit) is regulated, most operated by public bodies, there is considerable evidence of people's (poor) perception of transit services. Indeed, outside New York, transit is a minor mode of transport. New York has 20% of all transit trips in the US and 66% of all rail trips (in 2007). There is, however, evidence that in the US people can differentiate between bus and rail services and when given a chance will vote for higher local taxes to improve rail transit.

Recently in Houston, the Transportation Board wanted to substitute buses in place of approved new light rail lines. Voter pressure confirmed the light rail option, largely based on the first 7.5-mile line opened in 2004, which by

Table 2.1 Modal share in London

District	*Public transport*	*Private car*	*Walk/cycle*	*Total*
Central	20%	5%	75%	0.8 million
Inner	20%	30%	50%	4.1 million
Outer	15%	55%	30%	8.6 million

Source: Transport for London reports

2006 had reached the 2020 ridership forecast, carrying more than 45,000 passengers a day and reaching 40 million trips by September 2007.

In the UK by contrast, the local transit market has been deregulated since 1986. In many of the larger cities, some of the busier routes have two or more competing (bus) operators. There have been examples of urban bus services competing with rail lines, but none where rail lost out. In Manchester, when the Metrolink light rail system opened in 1992, one of the local bus companies ran express buses from Bury into the city center at lower fares than rail. Demand did not support the express bus service, which was withdrawn after 3 months. Similarly, in Sheffield when the Supertram light rail system opened to Meadowhall in 1995, an express service using articulated buses was operated with a fare half that of the tramway. It lasted 6 months before being withdrawn. People will get out of cars to make some local trips by rail transport but not by bus in most circumstances.

Recently there has been interest in guided busways, started in Essen (1980) and then copied with a longer line in Adelaide (1986) in Australia (Figure 2.2). This has led in the UK, after short-section experiments in Ipswich and Leeds, to new curb-guided busways in Edinburgh (CERT/WEBS), Leeds, and Bradford. In Leeds, the first curb-guided busway line on Scott Hall Road was reported to have increased bus ridership by 50%. On examination, most of this came from parallel bus routes not using the guideway. Transfers from cars were a statistically insignificant 3%.

Experience in Adelaide shows that even with a dedicated busway, many suburban branches cannot generate enough demand to support a through bus service, one of the claimed advantages of the curb-guided busway bus rapid transit. Instead, there are busway feeder buses, with passengers interchanging

Figure 2.2 Adelaide curb-guided busway (Photo: Virginia Transit Association)

onto through buses at interchange stations. This makes light rail equally practical.

Ottawa (Canada) has developed a busway network since 1983, covering most of the city. In 2007, the city council announced that it was considering light rail conversions for the future. This was after it was demonstrated that a short diesel light rail line attracted 10,000 riders per day. Construction of the first line of the new light rail system, using some of the busway rights-of-way, is expected to be completed in 2012. This surely indicates the limited market penetration that can be achieved by buses. Similarly, the recent abandonment of part of the pioneering Runcorn Busway reinforces the point.

More complex and interesting experiences have been reported from the new French light rail systems. Nantes was the first of the new-generation tramways opened in 1985. Two lines were built to replace the two busiest bus routes, and the bus network was reorganized to complement the new tramway. When it opened, there were timed connections at interchange stations and through ticketing. A third line was started in 2007.

Not only did the tram lines attract significantly more patronage (more than 50%) than the bus lines replaced, much of which came from diverted car trips, but the remaining bus lines also experienced an increase in ridership of about 20%. Part of this was explained by new journeys continuing by tram. By 2000, the first two tram lines were carrying 40% of all public transport trips in the city. What this proves is the importance of good interchange stations, between bus and rail and park-and-ride as foci for the new transport system if a significant modal switch from car to transit is to be achieved.

2.2 Light Rail Incremental Development

Historically, tramways and streetcar lines were built to tap the heaviest traffic corridors. The patronage attracted generated the profits needed to support later secondary and tertiary lines. These could be operated on a marginal basis by sharing core rail networks and central facilities like depots, management, maintenance, and other specialist teams. Two lines with interchanges were found to attract more patronage than two separate lines. The more complex the network, incrementally the more significant is the interchanging ridership, through "network synergy," especially when transfer or through tickets are available. This is on the basis that interchange is physically easy (for example, cross-platform) and the time and cost penalties are small (e.g., from high-frequency service and no price penalty) by use of, for example, unlimited travel tickets or transfers.

Nineteenth century streetcar and tramway systems encouraged middle-income families to move from the central area to the growing suburbs, which were served by the first-generation lines. During the 20th century, some lines were extended and others abandoned to be able to serve the growing suburban market, making better use of limited fleets. This reorganization continued until the Second World War, during which most systems enjoyed record ridership due to the shortage and rationing of fuel for cars and the mass employment of women in war work. Indeed, in many places tramways were extended to war work sites.

After the Second World War, most systems in Europe were at best badly run down and in need of urgent repairs and maintenance or at worst badly damaged and inoperable due to bombing. In the UK, bomb damage led to the closure of systems in Bristol and Coventry in 1941. In Germany after 1945, getting the local tramway running again was seen as the first sign of economic recovery, and it was the tramway, not buses, that got the investment, because of the scarcity of oil and rubber.

The strategy for operating networks also points out two contrasting approaches. In the US and UK, many cities have bus services that operate at low frequency, typically every 20 or 30 minutes. These attract only the transit dependent, who have a low value of time. In the US, many cities also have central business district timed interchange stations to allow "many-to-many" trips to be made with only one interchange. In the UK, low-frequency complex networks with direct "many-to-many" routes are the norm. These are used mostly by those without cars, since the long waits and timetable knowledge required are not acceptable to car drivers, who want the convenience of traveling at any time.

In many other developed countries, however, coarser networks are provided on a few-to-few basis. These lines have high service frequency, for example about every 5 minutes. Purpose-designed interchanges minimize the waiting time penalty, making journeys faster than would be the case with infrequent but direct (bus) service.

Clearly, a law of diminishing returns sets in as more lines are added. Historically, an optimum network maximizes ridership by serving the largest possible walk-and-ride catchment. This dynamic of urban travel was broken by the arrival of motor cars, allowing wealthier residents to live away from transit lines and to have a choice of travel mode by park-and-ride. This can be achieved informally by parking on side streets near stations or at purpose-built park-and-ride stations.

2.3 Abandonment and Reinvention

The first-generation streetcars and tramways did not always adapt to the changing market, where more people had cars and were less dependent on transit systems. Many systems merely retrenched, abandoning progressively the less intensely used lines. This put more central and overhead costs on the remaining lines. These then became less viable, since overhead costs could not be reduced as quickly as lines were abandoned and ridership reduced. This abandonment process in Britain, France, and North America lasted about 10 years, and with the network doomed, a patch-and-repair policy for remaining lines made them progressively more run down and less attractive.

It is also worth considering alternative approaches (Cudahy 1995). Many municipal streetcar systems could have been managed without the sudden abandonment they faced if a more businesslike approach had been adopted. Where managers saw a line operating at a loss, politicians saw a vital public service for the poor (voters). In contrast, managers would have chosen selective cuts in service to reduce costs and inflation-matching fare increases to maintain revenue. New contracts would have been negotiated with the unions to improve staff productivity. City hall usually turned these down without offering an alternative, until it was too late and the system could not be saved economically. Management struggled on but was unable to make any constructive alternative work.

When these abandonments were undertaken by public bodies, the resource implications of this approach should not be ignored. Private American streetcar systems were taken over by the National Transit Corporation in the 1940s, which was not much interested in transit as a business. This was because it was backed by large corporations that sold automobiles, fuel, and tires. Thus, running down streetcar lines pushed passengers to buying cars, from which the shareholding corporations made more profit, compared to writing off the capital value of the streetcar system.

Public authority abandonments were effectively destroying capital assets and in many places also scrapping nearly new vehicles, as in Ottawa in 1959. The recently promoted Merseytram line in Liverpool was costed at £360 million for its 19-km length. In 1957, the original tramway network was closed by a municipal decision made in 1949. This 400-km system at today's prices was worth £6.3 billion. This example can be duplicated in many European and North American cities. Such a large capital destruction can only be compared to the impact of war.

It is, however, ironic that in some North American cities, such as Boston, Baltimore, Philadelphia, Pittsburgh, San Francisco, and Toronto, small networks survived as a result of citizen pressure. In the UK, Blackpool continued to operate long after the last city tramway closed in Glasgow (1962). Similarly, a few short lines survived in France in Marseilles, Lille, and Lyon, before new systems were opened starting in 1985 in cities that once had tramways.

The picture in the rest of Europe was very different. Here, labor and imported oil shortages in the 1950s and 1960s made tramways the only practical option for urban mass transport. Reflecting passenger preferences, new rolling stock was purchased, which improved staff productivity and lowered costs. Extensions to new districts were built, even as inner city routes

Figure 2.3 Calgary light rail (Photo: Calgary Transit)

were thinned. This laid the foundation for continued service and operations and a growing importance as an environmentally sustainable public transport alternative to cars. Now, ironically, mainland European countries have higher levels of per capita car ownership than the UK but lower levels of car usage and higher public transport usage.

In the US, where per capita car ownership is about 10 years ahead of Europe, the need for an alternative to car commuting led to the opening of the first modern light rail system in 1981 in San Diego, which still holds the record of only 36 months between the decision in principle and opening for service. Calgary, though, was the first new system in Canada (Figure 2.3); its light rail system, which opened in 1981, also used the proven high-floor Frankfurt U2 car design from Germany, which was used in San Diego and afterwards in other cities. Since then, more than 20 new systems have been built in North America, many funded by local (fuel or sales) tax increases approved by referenda. Why do citizens vote to increase their taxes to fund light rail lines? Usually it boils down to the view that the light rail will attract enough car commuters to allow the rest to drive in less congested conditions. Table 2.2 lists some of the new systems recently opened.

News of the success of new light rail systems in North America soon spread across the Atlantic, where cities were fighting losing battles between rising car use and the decline of bus ridership. Experiments to improve city bus lines with reserved lanes, and in Runcorn a dedicated busway, showed little result in retaining riders, let alone attracting car users. More draconian measures to reduce congestion, like the removal of on-street car parking and

Table 2.2 Recent European light rail openings

Year	*Location*	*City population*	*City area (km²)*	*Population per km²*
1983	Utrecht, Netherlands	288,535	99.32	2,905
1987	Grenoble, France	157,900	18.13	8,709
1991	Lausanne, Switzerland	128,302	41.37	3,101
1992	Karlsruhe, Germany	285,812	173.46	1,648
1994	Rouen, France	106,592	21.38	4,986
1996	Oberhausen, Germany	218,898	77	2,843
1997	Saarbrücken, Germany	180,515	167	1,080
1999	Kaiserslautern, Germany	98,044	139.7	702
2000	Orleans, France	113,126	27.48	4,117
2000	Montpellier, France	244,700	56.9	4,300
2001	Messina, Italy	247,593	211	1,173
2006	Valenciennes, France	45,000		

Data collected by Colin Griffin, Galway Technology Institute

the introduction of parking charges, did not deter car commuting. More drastic measures to ban or tax car traffic consistently have been rejected by voters. Radical measures were therefore needed. This led to the consideration of new tramways, now called "light rail," in many cities which had only recently abandoned systems in favor of the modern and flexible bus.

The first reintroduction of light rail in Europe was in France, starting with Nantes in 1985; since then, new light rail systems have opened in Lille, Lyon, Bordeaux, Caen, Paris, Nancy, and Strasbourg. Like the new North American systems, the first new tramways in France were operated with high-floor vehicles, requiring platforms some 900 mm high to allow level entry for passengers with disabilities. Later systems have adopted low-floor vehicles, to make platforms less obtrusive and easier to install in central areas. Since then, many new lines and a few new systems have been built, some in quite small urban areas.

The first modern tramway in the UK was the Manchester Metrolink (1992) (high floor), followed by low-floor systems in Sheffield (1995), Croydon (1999), Birmingham (2003), and Nottingham (2005) (Figure 2.4). At this level of investment in light rail in the UK, it will take 150 years to reach parity with German city systems in 2004.

2.4 Alignment and Locations

The critical design objective for a new light rail line is the function it has to fulfill. As the UK National Audit Office reported in 2004, too often UK

Figure 2.4 Nottingham NET tram (Photo: L. Lesley)

systems have confused objectives, including urban regeneration and reducing social exclusion. The simplest objective is to maximize the diversion of existing car trips. Alignments are then easy to identify from the roads with the largest volumes of car traffic, both peak and all day. Such busy roads usually serve the most prosperous parts of the city, with the highest level of employment and car ownership.

The next stage is an analysis of the origins and destinations of cars trips, to determine those internal and external to the urban area. This is important, since the location of terminals, park-and-ride, and intermediate stations is critical to maximize the ridership a new light rail service may offer, from park-and-ride, walk-and-ride, maybe bus-and-ride, and bike-and-ride, at intermediate stations. The optimal conditions for siting park-and-ride stations are well established in terms of the distance from the final destination, location to the main access road, and size of the facility (Hovell et al. 1975).

Main roads can also offer rights-of-way for new light rail lines. If the objective is to maximize car transfer, then merely diverting some car trips to light rail and allowing suppressed car trips to fill the vacated road space is not a sensible option. In any case, a road traffic lane on a freeway can at most only carry 1700 passengers per hour by car at usual peak hour occupancy. The same space used for light rail can carry 6000 or more passengers per hour, an effective increase in passenger capacity of 300%. This is a strong argument for the exclusive use of some existing highway capacity for light rail, with suitable physical measures to ensure the uncongested movement of light rail vehicles. The main argument is that total passenger capacity is increased at a much lower cost than widening the road and adding more car parking. This of course ignores the environmental damage from extensive demolition to create a wider right-of-way and from the air and noise pollution of increasing motor traffic.

There are many examples of light rail lines operating in the median of major highways and several very satisfactory ways to introduce the rail line, with access to stations and stops in existing streets. The most important aspect is a proper partnership between the highway authority and light rail operator, to ensure that the light rail system achieves as near as possible 100% priority over road traffic. Light rail vehicles passing lines of car traffic in the peak period is a very visible reminder of the benefits of parking and riding.

Examples of possible light rail alignments are shown in Figure 2.5. Many urban areas also have little-used or abandoned railway lines. It is often a temptation to view these as ideal alignments, as they are quite straight and usually grade separated from the road network. Before too much time is spent on designing old rail alignments, it is worth considering why they are little used or closed. Many such lines were built for rail freight services, in the days when most goods were delivered by rail. Sometimes they were passenger lines but were abandoned because they were not in the right place to serve the present population distribution or did not give good access to important traffic attractors like the city center or local district centers.

Perhaps the original rail service was slow or infrequent. Where old railway lines do not fail for any of the above reasons, clearly they should be seriously considered and work expended to design accessible stations or stops, to attract

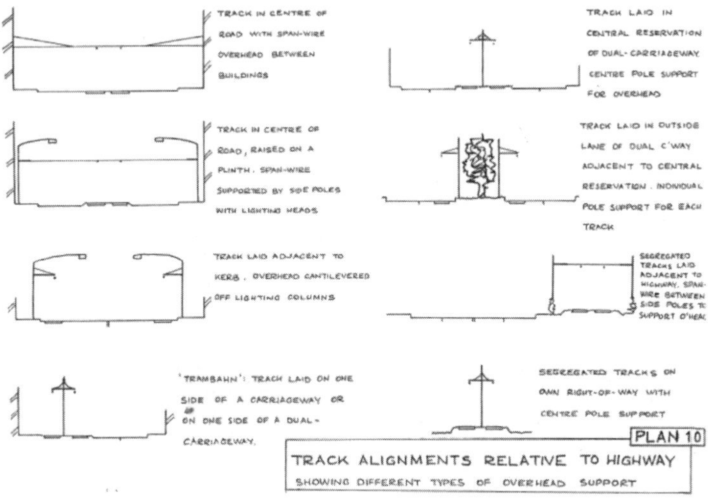

Figure 2.5 Possible highway alignments for light rail (Drawing: D.S. Hellewell)

Figure 2.6 Skokie Swift (Photo: Chicago Transit)

people to use the service. There is no doubt, as demonstrated by the Manchester Metrolink, Skokie Swift (Figure 2.6), and Tyne and Wear Metro, that replacing a conventional but infrequent suburban railway with a frequent and fast light rail service will increase patronage significantly compared to conventional main line rail service and much more than bus service. The option of replacing rail line in Tyne and Wear with busways was considered and rejected, because the capital costs were similar but light rail attracted more car trips.

Good access to the light rail system in densely developed areas like the city center means that for an affordable system, the tracks will be on the surface, mainly in existing roads, and possibly shared with other vehicles like buses. This is manageable if suitable traffic management measures are in force to allow light rail vehicles to operate in free-flowing traffic. The major benefit of such city center alignments is that stops can be relatively close together, to better serve the major traffic generators and attractors.

The second benefit is that installing tracks in streets (Figure 2.5) is much less disruptive and expensive than tunneling, which itself restricts the number of stations. There is also evidence of a considerable and growing resistance against using underground transit systems. Personal security is often cited as the reason. Many underground systems are also quite claustrophobic. Because they are on the surface, light rail lines are a permanent reminder of the service, even to those not using the system, and can be an important part of the civic fabric. Passengers waiting at a stop for a light rail vehicle feel more secure should a problem arise, by being able to summon help quickly or to escape in several directions.

If there is an argument for city center tunnels, it is to divert car traffic, most of which is traveling nonstop across the downtown. The controversial cross-city tunnel in Boston is an example of this. Most European cities divide the center into cells, with free movement from outside into individual cells but restricted car movement between cells, to reduce car traffic volume in the city center to only those that need access. This was pioneered in Gothenburg in 1975. Cross-center traffic is either diverted or restrained. This also benefits light rail transit, as less car traffic means less congestion and better service performance. The next stage up from this was pioneered in Zurich, where public transport services have automatic priority at all intersections with traffic signals. This has also been found to benefit car drivers, by marginally reducing their waiting times.

2.5 Operation

The starting point for a reliable light rail is to make the operation as simple as possible. This is so that:

- The system is easy to understand and use by passengers, thus maximizing patronage
- Capital and operating costs are minimized
- The system is easy to manage in real time, especially during unexpected events
- The system is less vulnerable to an equipment failure (e.g., by minimizing the need for signaling)

Historically, light rail has "run on sight," where the responsibility for safety rests with the driver, who has to drive like any other road user, so that the light rail car can be safely stopped within the visual distance. The driver also has to obey road traffic signs and traffic signals, which may have special phases for light rail to pass through junctions without stopping. Driving on sight on public highways requires special training in defensive techniques to avoid accidents. This is especially true for the use of emergency braking, which could injure passengers inside the light rail vehicle. In some cities, light rail vehicles may be afforded absolute priority over all traffic, except emergency services. Nevertheless, drivers must be aware of other road users and be prepared to slow down or stop to avoid accidents.

For a fully segregated system, faster operating speeds are possible and even full automation, like the Morgantown PRT in Virginia, Skytrain in Vancouver, the Docklands Light Railway in London, VAL in Lille, the Sky Line in Kuala Lumpur, and many airport distributor systems. Full automation is not, however, a low-cost option, in terms of either control equipment or the investment in the right-of-way infrastructure. It also requires highly trained technicians to keep the system running, whose salaries in total may differ little from a larger number of drivers not needed. Automation does, however, give flexibility in terms of running extra services at short notice for special events like major sports matches and to increase capacity when there is a short-term unexpected increase in ridership (e.g., during heavy rain). Automation has not yet developed to a stage where it can be used on public highways in mixed traffic.

Driver operation, running on sight, also means that light rail lines can penetrate the downtown on the surface, to maximize accessibility. This maximizes ridership to major attractors like shopping malls, office centers, etc., without the need for expensive infrastructure, as automated systems require. Similarly, the cost of closer stations is much less than required for subway systems. For a one-line starter system, the route structure will be relatively simple, with the exception of extra services operating only on part of the line. This allows more capacity to be provided near the central city core or where the demand is highest.

As a simple rule of thumb for transit services to be attractive, the waiting time should be less than the ride time (Cudahy 1995). If the longest ride is 30 minutes, then a 10-minute frequency would be acceptable. This means that riders at the outer end of lines will accept a lower frequency than those nearer the center taking shorter rides. In an urban environment, the outer suburbs may accept a 15-minute frequency. Obviously, the shorter the frequency, the greater the ridership that will be generated, as waiting time and therefore the total journey time are reduced. The feedback between these two is discussed in Section 3.2.

There is, however, a link between operating speed and frequency, when the fleet size is fixed. The higher the operating speed, the higher the frequency of service that can be offered for a given fleet size:

$$F = \frac{T}{n} \tag{2.1}$$

where F = service frequency (minutes), T = round trip time including terminal time (minutes), and n = number of light rail vehicles available for the service.

Higher operating speed also means that the light rail vehicles and staff are more productive, making more revenue journeys in a day. Raising operating speed is, however, complex, as there are interactions between the following variables:

- Stop spacing
- Maximum speed allowable
- Service acceleration and braking rates
- Traffic signals, priority, and preemption

- Time at stops to load and unload passengers
- Structure of the network and interoperation with other lines
- Service frequency

Each of these variables can be internally optimized and then will have an impact on the operating speed and hence system operating costs.

The operating speed also has an impact on the quality of service to passengers. The way passengers assess the overall service package can be determined by the use of generalized cost (Section 3.2.2). Ride time is one of the key variables in generalized costs. Generally, the shorter the ride time, the more riders, and hence revenue, that will be attracted. Therefore, for affordable light rail systems, the operating regime, and the management, should concentrate on maximizing the operating speed. A higher operating speed has an important impact on reducing or minimizing costs and maximizing revenue. Faster means fewer vehicles and drivers and thus reduces costs. A faster operating speed reduces passenger journey time and hence increases ridership and revenue.

When a new system opens, there will be a period of learning for the staff and passengers. This is true even if there has been a commissioning and training time before passenger service starts. As a result of this learning, drivers will become smarter, and scheduled service speed can be adjusted, allowing improvements in operating frequency and therefore being able to increase capacity to reflect growing ridership. Keeping the service comfortable for passengers, with a minimum of overcrowding, will be a key management tactic to maintain and to increase ridership and revenue.

2.6 Equipment and Standards

Light rail is not a proprietary system. There are many manufacturers of equipment that can be used to provide systems and services. This allows the light rail developer of a new system to gain competitive advantages by mixing equipment from a variety of manufacturers. There is, however, a downside in that the developer has to ensure that the different equipment is compatible.

There are also a number of standards which have been developed to assist both the manufacturers of equipment and system developers. Often these standards, like the German BöStrab, have been driven by operators that want to ensure that new equipment is compatible and can operate alongside existing equipment. If the design standards are performance related and specified, like the BöStrab, then manufacturers have considerable scope and incentive for product innovation. If, however, the standards are prescriptive with technical designs specified, this makes designs conservative and innovation difficult to achieve and will over time reduce the number of potential suppliers, since there are other industries that are innovative and seeking innovative equipment.

Light rail operates in a transport market that is dominated by the private motor car. Automobile manufacturers will upgrade models nearly annually and launch new models about every 5 years. With the 2009 energy and financial crisis, manufacturers will also update their technologies to make cars less polluting, more energy efficient, and cheaper to run.

Light rail vehicles designed for an operating life of at least 30 years will be obsolete in market terms long before they are worn out or technically obsolete. The first generation of streetcars built at the turn of the 20th century was by the late 1920s facing mass-produced motor cars, which had already been through nearly 10 generations of development and innovation, including all-metal bodies, heating, and upholstered seating. When compared to wooden-body streetcars with wooden seats and no heating in winter, no wonder the presidents of the privately owned US streetcar systems got together in the late 1920s to develop an answer to automobile competition. The Presidents' Conference Committee car (Figure 2.7) was the result. The first operated in Brooklyn in 1936 and then it was rapidly introduced throughout

Figure 2.7 Presidents' Conference Committee car in Newark subway (Photo: Courtesy Newark Transit)

North America. It gave the automobile a run for the money, as it was fast with smooth acceleration, comfortable seating, big windows, and heating in winter.

Light rail in the near future is unlikely to need the volume of equipment that makes mass production viable and therefore reduces costs. This is, for example, how the automobile industry produced mass consumer products, but it seems unlikely for a mass-produced light rail industry. The only exception was during the planned economy of the Soviet bloc, when the Tatra Company in Czechoslovakia between 1950 and 1989 built annually, on average, 1000 light rail vehicles based on the Presidents' Conference Committee design. These vehicles supplied virtually all the countries of the Soviet bloc.

Small production volumes dictate high prices, which in turn result in small orders, becoming a "Catch-22." At the time of this writing, a 200-passenger light rail vehicle costs about US$5 million, whereas a bus with the same capacity costs about $1.5 million. High first costs force (new) system developers to assume long life of operation but condemn such systems to becoming rapidly out of date in market terms. Mid-life refurbishment can however refresh the equipment, change appearances, and keep up with passenger expectations.

In any case, even if there were significant innovations that reduced costs or improved performance, because existing fleets have such long lives, market penetration will be very slow. Few operators can afford the luxury of disposing of a fleet of light rail vehicles only 10 years old, to buy new vehicles. The exception is Centro in Birmingham, UK, where a fleet of light rail vehicles with high maintenance costs is planned for scrapping after only 10 years of operation. Most systems are like Stagecoach, which operates the Supertram system in Sheffield, UK, where after 10 years of service the fleet of 25 light rail vehicles is getting a new livery and seating, at a total cost of £2 million or about 5% of the original tramcar purchase price.

In larger and more mature light rail networks, the rolling stock and equipment will be replaced on a 5-year cycle or in some cases annually. New equipment will be assigned to the busiest (most profitable) routes and older equipment cascaded across the network, allowing the oldest equipment to be retired. This at least means that all routes and passengers will have "new" equipment on a regular basis, allowing for continual passenger interest. The oldest equipment is then sold secondhand to another system, reused for maintenance vehicles, or scrapped, depending on market conditions.

A light rail developer of a new system therefore faces a difficult task, especially if the network is unlikely to be expanded in the future. In some 30 years time, the equipment will have to be replaced. How can the replacement capital be accumulated, especially if the original costs were granted from public funds?

Many new light rail systems may thus present huge problems to future management, when equipment begins to wear out, or become life expired, and thus needs replacing. If the system operates on a commercial basis, with a full depreciation account being built up, this will fund the replacement equipment. More likely, a publicly run system will, unless ridership exceeds projections, be continually cash strapped, and future equipment replacement will be a political issue and might, as earlier, lead to closure should public funds be scarce.

One way that this might be managed would be to divide the fleet in two; one part would be used for all-day every-day schedules and the other only for peak hours. The intensely used light rail vehicles will wear out first and can be replaced in less than 30 years, while the vehicles used less can run longer. This means that at least part of the fleet must be replaced in 20 years and the rest in 30 years, so the demand for capital expenditure is minimized and spread over a number of years. Should ridership increase, new light rail vehicles can be bought for both replacement and enhancement, allowing larger orders at lower costs to be purchased. An example of this can be found in Sheffield, UK, where shortly after opening it became clear that the fleet of 25 light rail vehicles would make operations difficult. A request for quotes for extra identical light rail vehicles resulted in a bid of £4 million ($6.5 million) for one and £10 million ($16 million) for four additional cars.

2.7 Understreet Utilities and Plant

When the first light rail systems, streetcars, and tramways were laid down in the 19th century, they were often the first utilities to be laid in streets. Subsequently, sewers, water supplies, gas, electricity ducts, telecommunication lines, etc. have used the same streets as a convenient route to service adjacent and frontage property. In those cities with a continuity of light rail service, a modus vivendi between the light rail operator and utility companies is in place. This is a variant of the "live and let live" philosophy. An example is stray currents, which are alleged to attack understreet utility plant and the steel frames of adjacent buildings and other structures.

In North America and most continental European countries where light rail has operated continuously for more than 100 years, no special measures are taken against stray currents. Where there is proof that it is the light rail system stray current that is causing local or specific corrosion problems, then the operator can install technical measures to protect those structures. Such measures can include the use of diodes or cathodic protection. As recently reported in the specialist press, one senior engineer (Snowdon 2007) observed that there has never been any scientific investigation of the magnitude or behavior of stray currents, let alone the effectiveness of the countermeasures installed in the UK, which he thought to be largely wasted.

These UK measures include the use of a "Faraday cage" under the track to "catch" stray currents. Not only do such measures increase the first cost of track construction, but maintenance has to be very careful to avoid damaging the protection and making it ineffectual. There are of course other ways to reduce stray currents. Whichever method is used, in periods of heavy rain, when the carriageway and tracks are awash with water, stray currents are unavoidable in a grounded light rail system.

New light rail system developers face utility companies with "grandfather rights" to the understreet space. There are two main reasons why understreet utility plant is relocated when new light rail systems are planned. One is a push factor and the other a pull. Typically light rail tracks are laid on a monolithic concrete slab about 20 ft (6000 mm) wide and 20 in. (500 mm) deep (Figure 2.8). Any utility plant in this area less than 20 in. (500 mm) below the road surface will be physically in the way of the track structure.

Utility plant deeper than 20 in. (500 mm) will however be inaccessible once the track slab is in place, and thus repair and maintenance will be impossible. Relocating such plant will at least provide access for the utility companies. Invariably, the replacement plant will use new equipment and therefore last longer than the residual life of the old plant. It may also have a greater capacity or improved performance. This "betterment" has been a source of contention for many years in Britain. Until 2002, the utility companies had to pay 18% of the cost of relocation of plant to reflect the

Figure 2.8 Manchester track slab, Market Street, 1990 (Photo: L. Lesley)

"betterment." Since then, the contribution has been reduced to 8.5%. Light rail promoters see this as a subsidy to utility companies, which often have old plant in city streets.

On the other hand, utility companies often prefer to leave plant *in situ*, since their own design offices are stretched in providing new plant and connections. Invariably, relocating old plant also creates secondary and connection problems. Many utility companies subscribe to the philosophy "if it isn't broken, don't fix it." The capital-conserving financial model of many utilities is to repair when needed and replace when it becomes more economic than continuous repairs.

The second reason for relocating plant is to avoid disrupting light rail services when utility companies need to make repairs or replacement. This relocation is a very costly luxury. When the 3.4 km (2.25 miles) of city center tracks in Manchester where laid in 1990/1991, the cost of utility relocation was £8 million ($13 million), while the cost of track laying was only £6 million ($10 million). Clearly, if there is a catastrophic utility failure like a burst gas or water main, then light rail services will be disrupted anyway. Such failures are, however, rare, about a one in 40-year event on any particular street. For routine maintenance and replacement, alternatives like temporary crossovers (Figure 2.9) with single track and two-way running, a short gap in operation with passengers walking or being bused, or utility companies undertaking work during the overnight closure of light rail service

Figure 2.9 Temporary track crossover, 1904 (Photo: Liverpool City Engineer)

are but a few ways in which the majority of understreet plant could be left in place, saving capital costs and construction time.

There are several other reasons why moving existing utility plant should be the last option. In many historic cities, the understreet space is already extremely congested with plant. Finding space for new plant is a major challenge. If utility plant diversion is undertaken, this is very disruptive to the economic life of the city, as the citizens of Edinburgh discovered in 2008–2010. One strategy is to divert utilities down backstreets and then feed utilities and ducts across adjacent properties. Getting this level of cooperation from property owners is extremely time consuming and difficult, as well as costly.

Another reason for leaving plant in place is that utility companies can barely keep up with routine repair and maintenance. Having to design the diversion of plant may also require major changes to network structures and cause downstream problems where new plant interfaces with old. Relocating utilities always requires deliberately breaking into old plant.

How can the light rail system developer minimize the cost of utility relocation? The first step is to have a meaningful consultation with the plant owners, so that utility problems can be fully understood by the light rail system designer. Often, very small realignments of track are all that is needed to avoid the costly moving of plant.

In one project, a small sideways realignment (40 in./1 m) of more than 164 ft (50 m) of track at the design stage avoided a £2 million relocation of the main 250-kV AC supply cable for a UK city. Similarly, often utilities can see the benefit of working with light rail promoters if a genuine atmosphere of cooperation has developed. In another project, one of the utility companies offered to issue ducts to the light rail developer free of charge, which could be put in the street during track construction. This would mean that in the future when new utility services were required, they could be fed down *in situ* ducts, without having to dig up the street and disturb tracks or operations. This level of cooperation has been agreed to for the new GLUAS light rail system in Galway, Ireland.

3

Planning Light Rail

3.1 Setting Goals and Objectives

All planning must start with a clear set of objectives. Since planning, including traffic and town planning, became an academic discipline, a considerable body of literature has been written on the importance of setting clear objectives in plan making (Figure 3.1). Confusing and sometimes mutually contradictory objectives were heavily criticized by the UK National Audit Office 2004 light rail report. This was a wide-ranging and severe criticism of the new light rail and tramway systems it reviewed, which had been open since 1990 in Britain.

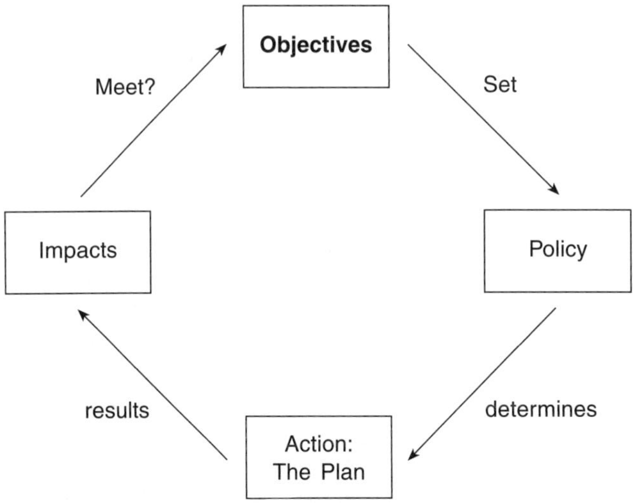

Figure 3.1 Plan and objectives (Drawing: L. Lesley)

In particular, the promoters of the systems had claimed objectives to justify the public funding that were rarely met in part, let alone in full.

The more objectives a project is trying to achieve, the less likely it is any of them will be satisfied. Often in a transport, or light rail, project there are implicit or hidden objectives, like "civic keeping up with the Joneses" or the ruling party rewarding the parts of the city where its electoral strength lies. In the US, this is called "pork barrel politics." Is seeking votes in marginal areas of the city close to gerrymandering?

What are suitable objectives for light rail and how can they be framed? There is a large body of planning literature on this topic, and this is an attempt at a concise summary. Objectives can be normative or derived. In the first case, the objective is to achieve an accepted norm, like reducing air pollution to the specified healthy or safe level or enabling people to reach their place of work within a defined time. Normative objectives usually can be measured and therefore can be readily confirmed as to the effectiveness of achievement by a project. For a light rail system, these would be carrying x% more passengers or achieving a y% market share.

Derived objectives begin with an analysis of the existing situation to identify particular problems found in that city. Often, subjectivity confuses such analyses. This certainly can be the case when comparing the severity of different attributes, even if it cannot be measured exactly. For example, is reducing traffic congestion more important than encouraging development? Derived objectives usually emerge from market research, public consultation exercises, or consultants' studies. Objectives like reducing traffic congestion or social exclusion fall into this category, since even defining what these "problems" are is difficult. Measuring them is also difficult. Identifying any statistically significant change, against what is often a noisy data background, can be impossible. This can degenerate into the kind of medieval debate about the number of angels that can fit on the head of a pin. Nevertheless, much political debate is undertaken using this kind of language or rhetoric, and without doubt, where public bodies are involved in promoting or funding light rail projects, derived objectives will be part of the justification for a new project.

Another approach is the identification of widely accepted problems and then seeking their solution. This approach can be more difficult, as there is rarely a consensus as to what constitutes a problem. On the other hand, if problems can be agreed upon, then solving them is easier than planning to achieve normative objectives. Solving problems might seem easier, but it is not so easy to achieve, since there is usually more than one solution that can be applied. The job of the (transport) planner is, then, to find, examine, evaluate, and select the alternative most likely to achieve a solution to the accepted problem. Doing this without creating further problems—the so-called law of unintended consequences—is not always so easy. Solving prob-

Figure 3.2 Park-and-ride and bus interchange at Dresden light rail transit (Photo: Dresden Transport)

lems and providing the best value for the money invested, from whatever source, is the most difficult challenge of all.

It ought to be obvious that transporting people is the most important objective for a light rail system. In most American and European cities, the car is the main means of personal transportation. The objective of a light rail system should therefore be to change the modal split, at least of peak period travel (Figure 3.2). Achieving a change in modal split is easy to measure and, given the size of the motor car market, provides the light rail system promoter with a large pool of trips from which to attract ridership and revenue.

A light rail system that does not carry passengers must surely be a non sequitur. As the UK National Audit Office revealed, this has nearly happened in the UK. Promoters claimed that the light rail lines built would break even on operating costs, but they have required considerable subsidy to be kept open, because patronage did not reach 50% of the promoters' forecast.

Simply measuring the number of riders using a new light rail line is not adequate and can be misleading. Riders may have diverted from parallel bus or rail services, not increasing total transit ridership. Establishing that riders have genuinely been attracted from private car use is also not without problems. Before and after studies of household travel behavior can identify statistically significant modal split changes. These can be confirmed from transit rider surveys. Similarly, cordon surveys of commuters entering the central business district can identify modal shifts, especially where linked to long-term passenger and vehicle counts on the corridor benefitting from the new light rail line.

Where new light rail projects are promoted and in part funded by property developers, the rise in land values or property prices will be of considerable interest to the developers, who will wish to see their investment in light rail rewarded. Indeed, a pioneering study of the Yonge Street Subway Line, which opened in 1954 in Toronto, showed that within 5 years much of this transit investment had been recouped from the rise in land values and sales around station locations (Dewees 1976). Similar conclusions were identified more recently in Seoul (Bae et al. 2003).

The first stage of the London Docklands light railway, which opened in 1988, was paid for within 3 months of the start of construction, through the rise in the value of brownfield sites and land sales. In many places, however, such overheated land value conditions do not apply. Declining heavy industrial cities may effectively have land that has a negative value, due to contamination and the cost of restitution, especially if the population is also in decline.

Unfortunately light rail will make little difference in such cases, as witnessed in the Midland Metro in Birmingham, UK, which crosses large areas of industrial dereliction around Wednesbury ("The Black Country"). The lack of development therefore means it does not carry enough passengers to cover costs. These are cross-subsidized from National Expresses local bus services. Nor has this light rail line produced either brownfield redevelopment or economic regeneration.

This is an example of a political light rail project. A previous attempt to build a light rail line was rejected because of the large areas of demolition proposed. Ironically, the name of a citizens' action group, SMART (Smethwick Against Rapid Transit), is 'TRAMS" spelled backwards. The "soft" option of an abandoned rail right-of-way was chosen instead. This ignored the fact that it hardly served any residential or work areas but does serve a football stadium. The traffic from that however (one afternoon every two weeks) is not enough to compensate for the lack of traffic the rest of the time.

Other objectives like urban regeneration, improving the environment, or reducing social exclusion are unlikely to result in extra revenue for the light rail system promoter. These and other worthy social and similar objectives, while conferring community benefits, assuming they can be detected, are unlikely to assist directly in the finances of the light rail project.

Social objectives might be considered as fortunate side effects but really should not be counted as key. If they are, is this a sign of a weak project? Light rail, like transport in general, is a service industry and not a driver of the economy. Cities with weak economies need to address the economy and not imagine that light rail is a miracle regenerator. This is true even if a property developer is promoting the project as part of a package to open up access to an underused site. In this case, the property developer will pay for all or part of the capital cost and offset that against the rise in the value of his land bank that otherwise might not be developable. Economic regeneration came from the redevelopment of the land, not the construction of a light rail system. Light rail promoters must not confuse cause and effect.

3.1.1 Public Consultation

In North America and Europe, there is a legal requirement for public consultation for all publicly promoted projects, including light rail. Consultation is not just informing the public of plans. The reason for this is partly the democratic process, to ensure that public money is wisely spent. It is also partly a "good neighbor" exercise, to determine what the effects on residents and businesses in the area will be and how any negative effects can be ameliorated. There are formal consultation processes in a number of countries. In the US, these have been set out for more than 30 years (US Department of Transportation 1976) (Figure 3.3).

The key to effective public consultation is stimulating feedback, both positive (support) and negative (objections). Normally, people with objections are more than willing to make them known and, if there are enough, to band together into an opposition group. The promoter must then address

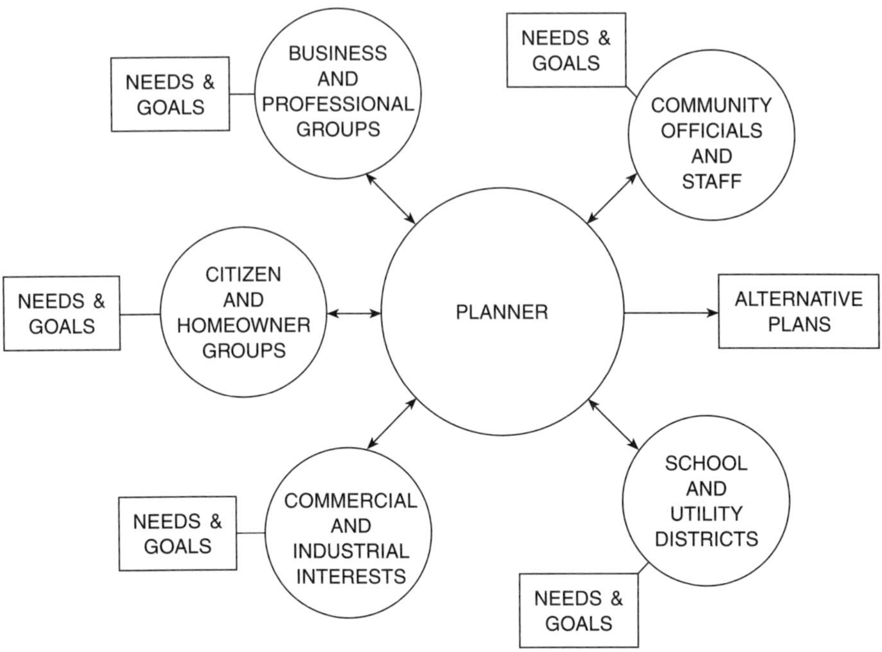

Figure 3.3 US Department of Transportation (1976) consultation guidance

fully all objections. There are two main ways to achieve that. In the first, the light rail plans can be modified to remove, or at least substantially resolve, the issue. If this is not practical, then the objectors can be compensated. While there are many examples where public consultation was successful, the cases where consultation has failed can also be salutary.

One such example of failure was the West London Tramway. It was promoted by Transport for London to replace tram line 7, originally abandoned in 1938 along the Uxbridge Road, replaced by electric trolleybuses, which in 1962 were themselves replaced by diesel engine Routemaster buses. The West London Tramway was to be built through the boroughs of Hammersmith, Ealing, and Hillingdon. Very quickly, residents and business organized an opposition campaign, based on a number of objections. The objections were mostly very reasonable and revolved around the residents' access to facilities in their area and the impacts on local businesses.

The campaign was led by a national journalist, Virginia Ironside, also a local resident. All the objections could have been resolved if measures and traffic management techniques in widespread use in the rest of Europe had been offered. Instead Transport for London took a "take it or leave it" approach. The result was that in the 2004 municipal elections, the ruling Labour Party lost control in Ealing and Hillingdon to the Conservatives, who had campaigned on an "anti-tram ticket." The Conservatives immediately withdrew support. The Labour-controlled Transport for London had no choice but to abandon the project, after spending £50 million on preparations and consultation.

For a new generation of privately funded projects, as well as the legal consultation process, the promoter will begin with local market research, to find out residents' views of their district and city. A private light rail promoter also will seek to influence public opinion. Champions will meet civic leaders to present the case for and the benefit of light rail to the city.

From whichever angle a new light rail system promoter comes, good public consultation is more than a legal requirement; it makes very good sense to ensure the system is acceptable to the citizens through whose "front yard" it will pass. As importantly, public acceptance is needed to gain political support and the maximum ridership. For the public sector, ridership should be the most important objective. For a private promoter, maximizing the revenue generated will be critical to achieving the business plan projections and satisfying investors.

3.2 Demand

How many passengers will a new light rail system carry, and what is the maximum revenue that can be generated? The answers to these two questions are not the same, because the fare elasticity for urban (public) transport is rarely –1.0; more normally, it is about –0.3. In one project with a 19-km light rail line, the maximum patronage was found when no fare was charged and a maximum revenue at about 65% of the maximum patronage. In urban areas, there are already people making trips. The purpose of those trips is known: to go to work or for education (the two most important economic activities) or to go shopping, visit friends, seek medical treatment, and so on.

Similarly, the method of transport used to make these trips is also known—or easily determined. Finally, as nearly 80% of trips start or finish at home, the origin and destination of trips, together with the start and finish time, can be determined. This data is vital in the planning of a new light rail line, as it allows modal shift calculations to be made and therefore the likely ridership that can be attracted and under what conditions.

Looking at data (e.g., 40 years of the UK National Travel Survey) that records the travel pattern of people making trips for all the above reasons, one fact stands out: the number of trips made per capita has remained remarkably stable at about 1100 annually. What has changed is the average trip length, reflecting the real decline in the cost of travel during this period and the modal switch to the use of private cars. Why has the number of trips made remained stable? The principal reason is that trips for work (or education) dominate most people's lives, and normally these accounts for two trips on most days.

In fact, the amount of travel consumed reflects the classic economic demand/cost curve (Figure 3.4). In the future, the number of trips made is

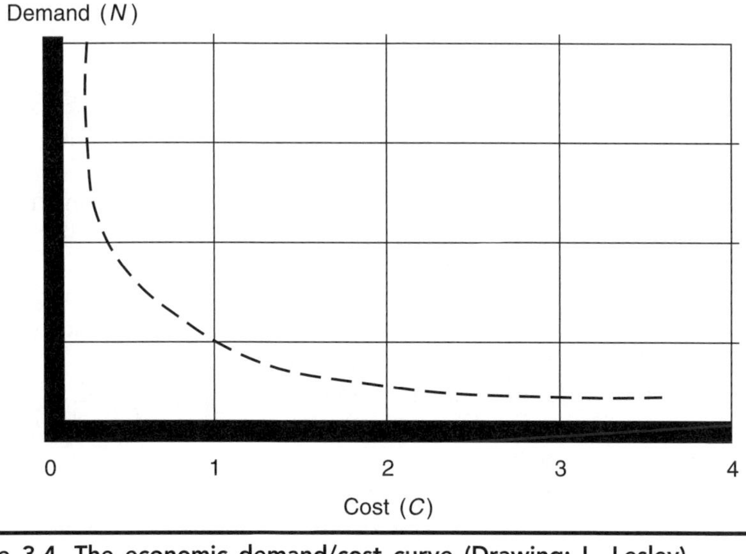

Figure 3.4 The economic demand/cost curve (Drawing: L. Lesley)

likely to remain broadly static, since there are only so many activities that people can undertake in a week. What will change are origins and destinations and therefore trip length and the mode of transport.

The light rail promoter relies on the fact that a person's mode of transport can be changed in the short term. In the light rail corridor, this can be immediate by people in cars switching to light rail. In the medium term, with the swirl of people changing jobs and housing, more will choose to locate for convenience to the light rail line. Also in the medium term, the light rail system, by enabling or perhaps providing more intensive economic developments, will also change origins and destinations, by allowing more people to live or work near the line.

As people move to live closer to the light rail line, this will enhance adjacent house prices. A 30% rise was reported from the opening of the Tramlink system in Croydon (UK). The enhancement of property values near light rail lines, compared to those further away, has been noticed in many cities. This urban restructuring can reinforce the role of the light rail system and capture a larger share of urban trip making, from typically 2% to more than 5% of all trips per line, or up to 40% of the movements in that corridor.

3.2.1 Origin and Destination Traffic

Ridership on a new light rail line will be attracted or diverted from the existing travel pattern. Conventionally, an urban area is divided into, as far as possible, homogeneous traffic zones (Figure 3.5). Travel between zones either will be determined from household or similar surveys (Bonsall and O'Flaherty 1997) or will be calculated from the population for origin traffic, and number of jobs and retail or leisure space for the destination.

Plotting the desire lines between all combinations of origin and destination zones produces a bewildering pattern for a metropolitan area of about one million people. For this reason, the data produced is recorded in an origin-destination matrix (Bonsall 1997). While data for trips by all modes and for all purposes can be recorded in a single matrix, the light rail planner will be particularly interested in car trips for all purposes, since these will represent the largest and most attractable traffic potential.

The light rail promoter will also need origin-destination data for trips between the zones that lie along the proposed light rail line or, the other way around, will seek the lines through zones which already have the heaviest traffic flows. Trips from one zone to another which passes through other zones en route will also be of interest. In designing the light rail alignment, the planner will seek to maximize the number of these trips that can be attracted by convenient stations and journey speed.

In most urban areas, there are few end-to-end trips on any particular light rail line. Most trips are between intermediate points. Therefore, knowing the

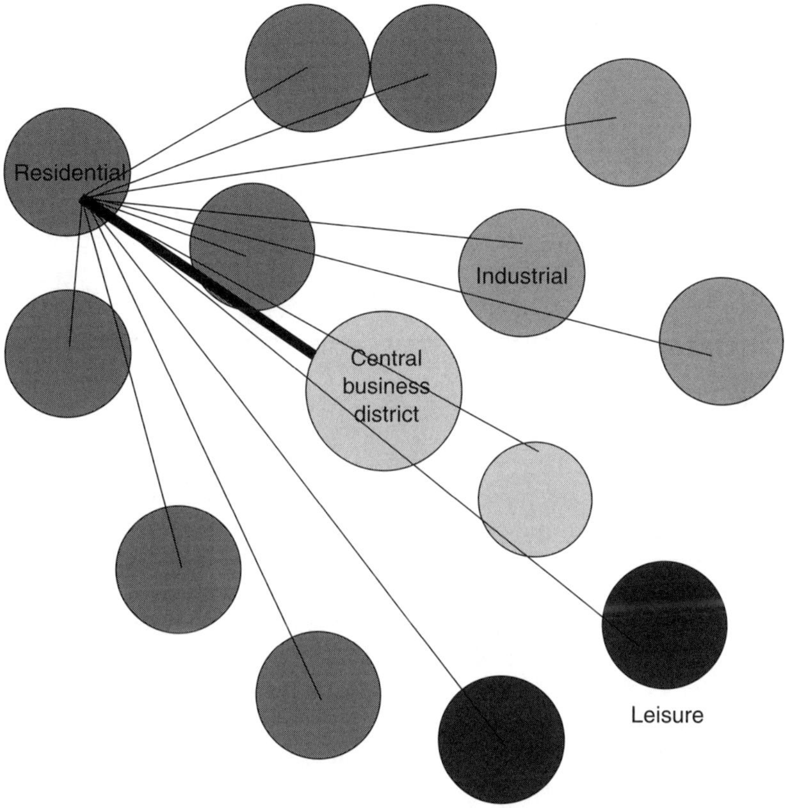

Figure 3.5 Urban traffic zones: desire lines from just one zone to all others (Drawing: L. Lesley)

origin-destination matrix data is an important foundation, both in designing the lines of a new light rail system and determining the ridership that can be attracted.

3.2.2 Generalized Cost

In economic theory, the demand for products (or services) increases as the cost declines and vice versa (Figure 3.4). This classical analysis is based on the cost as the actual cash transaction for obtaining the product or service. Over the last 30 years, transport economists have used considerable market research to demonstrate that the cash cost of transport is not a good measure of people's travel consumption behavior, which was already known due to fare elasticity.

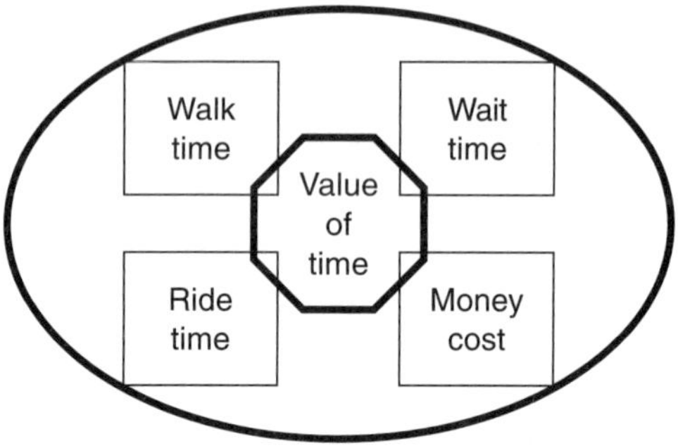

Figure 3.6 Generalized cost relationship (Drawing: L. Lesley)

A result of this market research is that other factors were found that influence choice of travel mode. These factors take into account the comparison between alternative modes of transport that can be used for a journey. Most importantly, the quality of the journey was considered more important than the financial cost of the travel. All of these factors were moderated by a traveler's value of time. Bringing these different factors together was made possible by the construction of a "generalized cost."

Generalized cost has a good correlation with passenger transport demand (Figure 3.6). Generalized cost brings together financial and service attributes and then weighs them against people's willingness to pay. This is usually a function of income and the economic importance of the journey. In broad terms, richer people are willing to spend more on transport to save travel time, by buying speed, while poorer people spend more time traveling on lower cost modes to save money. There are many volumes devoted to the consideration of generalized cost, but here we will consider its basic relationship. A fuller explanation, references, and worked examples can be found in Appendix 1.

The starting point is the fundamental relationship in the demand/cost curve:

$$N = \frac{k}{GC} \tag{3.1}$$

where N = the demand (measured in passenger kilometers), GC = generalized cost, and k = a constant, usually city specific and also a calibration constant.

This relationship is good for specific journey purposes, so a disaggregated analysis of each journey purpose should be undertaken to arrive at the most exact relationships needed for light rail traffic predictions. Fortunately, most of the journeys made in the peak period are work related, and a very acceptable approximation can be calculated by analyzing two journey categories: work travel and all other travel as nonwork.

The advantage of generalized cost is that it allows comparisons to be made between different modes of transport, as well as the impact of changes in one or more aspects of a particular mode.

The generalized cost function can be written as:

$$GC = c_1 \cdot T_1 + c_2 \cdot T_2 + c_3 \cdot T_3 + c_4 \cdot F + c_5 \cdot D \quad (3.2)$$

where c_1, c_2, c_3, c_4, and c_5 are constants and T_1 = walking time, T_2 = waiting time, T_3 = riding time, F = out-of-pocket costs (fare), and D = a measure of perceived comfort and safety. From a large number of studies, c_1 and c_2 are nearly equal to 2. Often, c_1 is about 1.8 and c_2 about 2.2, indicating the relative willingness of people to walk but a lower inclination to wait. c_3 usually is almost 1.0, meaning that people perceive riding time at about clock time. c_4 is about the inverse of wage rates for work journeys and between 0.5 and 0.1 for nonwork journeys.

People's disinclination to wait is nicely illustrated by a study undertaken by London Transport at Earl's Court Underground station in London. Here there are both elevators and escalators between the deep-level platforms and the street. On average, elevator time is significantly shorter than escalator time. Most passengers use the escalators, because normally they can be used without a wait.

Equation 3.2 produces a value of generalized cost in terms of equivalent minutes, as opposed to one based on equivalent money. Using equivalent minutes as the measure of generalized cost has a number of important advantages, since:

- Time is a measure that can be compared across different countries
- Time is inflation (money) proof, allowing direct comparison over a number of years
- Time intuitively gives a better measure in making a lower generalized cost correlate with increased travel
- Time allows different income groups to be considered on the same basis

Generalized cost can be used to determine the absolute measure of travel demand or, given existing data on demand, can be used to analyze the effect of changing the transport system as follows:

$$\frac{N_1}{N_2} = \frac{kGC_2}{kGC_1} \quad (3.3)$$

Therefore:

$$N_2 = N_1 \cdot \frac{GC_1}{GC_2} \qquad (3.4)$$

where N_1 is the existing travel demand, GC_1 is the existing generalized cost, GC_2 is the generalized cost of the new situation, and N_2 is the resultant travel demand.

This has the further advantage that the calibration constants cancel, assuming they remain the same or nearly the same during the period of change. Normally these calculations are undertaken for changes during the next year, rather than over 10 years. Thus, only the relative change in generalized cost needs to be determined, making predictions of changes in demand that much easier to calculate, including sensitivity analyses.

3.2.3 Elasticities

Another way to determine travel demand, although normally only on aggregated data, is the use of elasticities (Figure 3.7). This is the economic measure of people's willingness to pay and can be considered as the slope of the demand/cost curve at the particular point being considered. In the urban public transport industry, a fare elasticity of about –0.3 has been the norm for about 40 years. This means that if fares are raised by 10%, demand drops

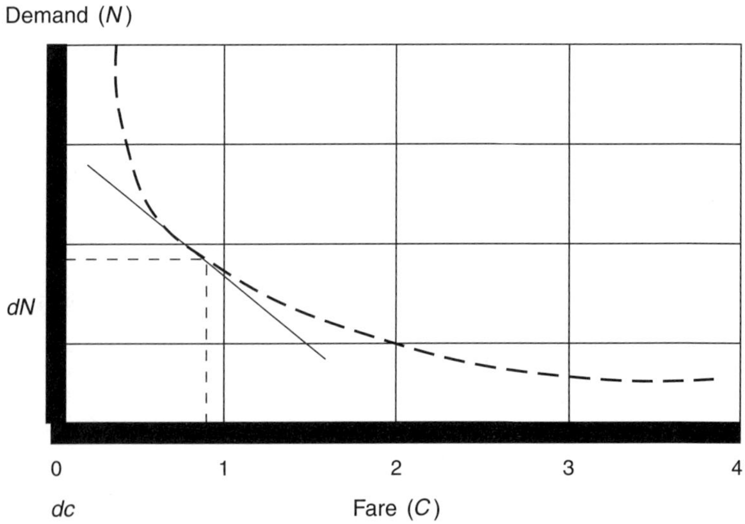

Figure 3.7 Fare elasticity (Drawing: L. Lesley)

by 3%. Total revenue increases by about 7%. This confirms the basic point of generalized cost. The cash/fare element in the decision to travel is not the only or even most important factor. The equation to calculate fare elasticity can be determined from the slope of the fare/demand curve (Figure 3.7):

$$e = \frac{dN}{dc} \tag{3.5}$$

where dN = the change in demand caused by the change in fare, dc = the change in fare, and e = elasticity of demand.

What it also means is that it has been assumed that fare elasticity behaves in the same way for fare reductions. Thus, a 10% fare decrease will lead to only a 3% patronage increase, leading to a reduction in revenue of about 8%. The literature on fare decreases is much smaller than that on fare increases, since there have been fewer of them, during a period when there has been constant inflation, and although fares have sometimes not risen with inflation, due to subsidies, this is not perceived to be a fare reduction. In fact, in most places fares have risen faster than inflation, to compensate for the systematic decline in (bus) traffic.

There are cases of fare increases leading to patronage increases, due to the change in the fare system. An example of this was in 1979 when the Rhein-Ruhr VV in West Germany changed from a route-based distance fare scale to a simple zonal fare system. This coincided with an average fare increase of 10%. A patronage decline of 3% had been budgeted. In fact, patronage rose by 4%, allowing for new investments that for the next 20 years created positive feedback on ridership and revenue.

Given the level of money inflation over the last 40 years, it has not been possible to distinguish with any confidence the long-run elasticity of fares. Few passengers have short-term options and cannot make immediate changes in travel behavior. In some countries, there are programs to encourage car sharing. Where this is possible, a transit user can team up with a car driver at work, share the fuel costs, and become transit independent.

The intuition is therefore that the long-run elasticity is higher than the short run. Passengers without immediate choices can over a period make radical changes to their lifestyle. Such changes include buying a car, moving one's home to be more convenient for work, and changing jobs as the three most obvious. These will have profound impacts on mode choice and the ridership of transit services.

Studies on the impact of new car purchases in the UK showed that the first car purchased in a household reduced public transport use by about 360 trips annually, whereas the second car purchased reduced transit use by about 280 trips a year (Hills 1981; Goodwin 1993; Paulley et al. 2004; Cheek 2008).

More recent work (Paulley et al. 2004) shows that urban bus fare elasticity in the UK is now about –0.4, reflecting the fact that many bus trips are optional, as very few people are dependent on buses for the work journey. The implication of this higher elasticity is that a 10% fare increase will boost revenue by about only 5%. Bus operators cannot therefore assume that rising costs can be met by raising fares, which has been the method used by most operators in the past 40 years. This has exploited the short-term fare elasticity while ignoring the long-term elasticity. Now that the short-run elasticity is higher, raising fares will merely collapse the business faster. This is the main reason why UK bus operators have recently concentrated upon cost reduction measures, providing useful experience for potential light rail promoters.

3.2.4 Prediction Accuracy

In planning a new light rail system, the most problematic of the predictions is the patronage forecast. Indeed, the UK National Audit Office 2004 light rail report singled out for criticism the poor accuracy of project ridership forecasts. This was in terms of either not achieving the predicted patronage at all, and therefore lower revenue than budgeted, or at best forecast ridership takes significantly longer to be achieved. Both of these therefore lead to an unbudgeted period of losses, which often are rolled up into the capital costs to be covered by the original grant or public funding.

Light rail promoters normally have few options for reducing costs when opening a new system, if patronage and revenue do not reach forecasts. The infrastructure is fixed and cannot be used for anything else. Some of the rolling stock can be mothballed and staff laid off. This, however, is likely to worsen the problem. A lower frequency of service will further reduce ridership, until the ultimate situation of no service, no revenue, and an idle capital asset.

One way to handle the inevitable, relative inaccuracy of patronage and revenue forecasts is by the use of sensitivity analysis (Table 3.1). Here, in turn, each variable used in the prediction is altered by the same amount, usually 10%, and the impact on the whole forecast is determined. From this, the variable(s) most critical to the forecast can be identified. This allows further work to refine the data for that variable and therefore improve the accuracy of the whole patronage forecast. Ranges of patronage forecasts can also be used in the economic appraisal model to see how sensitive it is to inaccuracies in the patronage forecasts.

None of these techniques by themselves will make patronage, and therefore revenue, forecasting 100% reliable, but at least the likely range of forecasts can be determined and narrowed. A business tool like mini-max analysis can also be used to minimize the worst-case scenario. In a commercial project, investors are likely to consider only the low estimate of ridership demand,

Table 3.1 Sensitivity analysis to different operating speeds

	Speed (km/hour)					
	20 base		*22 faster*		*25 fastest*	
Lines	*Passengers**	*Fares***	*Passengers**	*Fares***	*Passengers**	*Fares***
1	0.73	2.9	0.74	2.9	0.74	3.0
2	1.29	5.1	1.29	5.2	1.30	5.2
3	0.69	2.7	0.69	2.8	0.70	2.8
4	0.20	0.8	0.21	0.8	0.21	0.8
5	0.65	2.6	0.65	2.6	0.65	2.6
6	0.88	3.5	0.88	3.5	0.88	3.5
7	0.50	2.0	0.50	2.0	0.50	2.0

* Million passengers per year
** Million dollars per year

on the basis that if that is adequate to finance the capital, then should demand be higher, investors will enjoy comfortable profits. Publicly funded projects appear to have used only upper forecasts, if the UK National Audit Office conclusions are valid.

3.3 Performance

The performance of a new light rail system will be governed by a number of factors, including:

- Quality of the design of the route and stations
- Type of equipment and vehicles used
- Caliber and training of staff
- Operational management and timetable robustness

There are examples of apparently similar light rail systems that have very different performance in terms of passengers carried per vehicle or staff member and number of vehicle kilometers run per year or per staff member. These differences have a major impact on costs.

An example of this is the Supertram system operating in Sheffield, England (Figure 3.8). When set up, it was manned on the basis of six drivers per vehicle with off-board revenue collection from ticket machines. These proved unreliable and expensive to maintain and were frequently vandalized. Patronage built up more slowly than forecast, resulting in an operating loss, which the public authority was unable to fund. The operation was therefore

Figure 3.8 Sheffield Supertram (Photo: Paul Cantrell, Starquake)

privatized to the multinational company Stagecoach, which initially reallocated half the drivers to be fare collectors on the vehicles. Off-vehicle ticket machines were abandoned.

This immediately increased revenue by 7%, which paid for the wages of the fare collectors/conductors. Only three drivers per vehicle was very close to the Stagecoach overall average of 2.9 for its bus fleet. Increasing revenue and reducing costs meant that within 2 years a system that had been losing money was generating an operating profit of about £1 million a year, as noted by the UK National Audit Office, nearly the same sum that Stagecoach paid for the 20-year operating concession.

3.4 Stations and Stops

Passengers access light rail systems at stops or stations. These are the first physical points of contact with the service. They are the "shop window" of the system and will bring into perspective whatever marketing image has been created. Functional and aesthetically pleasing stations need not be expensive and because they have low running and maintenance costs can be a better investment than, say, additional lines or more vehicles.

Stops need to be located to allow convenient access to where people live, work, shop, attend school, etc. In a supply-side analysis, it has been assumed that people will walk 400 m to a transit stop, unless urban development is of varying density. This implies a stop spacing of between 400 and 800 m, with a continuous corridor 800 m wide served, for walk-and-ride passengers.

With park-and-ride, or bike-and-ride in some countries, stations can be further apart without reducing the effective population catchment but with a faster service speed and shorter travel time.

There is an interaction between stop spacing and operating speed. Fewer stops mean a higher operating speed but a smaller passenger catchment area. More stops slow the operating speed but shorten the time spent at each stop to let off or pick up passengers. Like the supermarket line, transit station stop time is a critical factor in maintaining service regularity. One of the main advantages of light rail is that stations have a relatively low capital cost and can be quickly built, even when in street. Having assessed the likely potential traffic generation at stops, a new light rail system might therefore concentrate on the half with the greatest traffic potential and be prepared to add new stations later as revenue is generated, to further increase patronage. This is a decision that can be made only on the basis of excellent local market intelligence.

Adding new stations later has several additional advantages. First, once a system is open, operating performance should improve and therefore higher operating speeds can be achieved, so new stations can be added without needing new rolling stock, by using the extra efficiency gained from experience to maintain the original operating speed. Second, in terms of the local travel market, it allows the operator to be viewed as developing the business, by opening another "outlet" for more customers. Third, it allows the operator to undertake joint ventures with property developers, on land adjacent to the line, and open new stations integrated with the development. Indeed, electric light rail can pass through buildings without any special environmental protection, as is the case in Ludwigshafen, Germany, where a department store was developed with a tram stop inside (Figure 3.9).

As the light rail stations are the shop window of the system, it makes good sense to have high-quality designs and finishes. It is a strategic decision whether to have the same design for all stations or a local design for each station to reflect the personality of the neighborhood in which it is located. Whichever approach is adopted, a corporate identity and clear signage will help passengers find and be guided around the light rail system. Good design should help to reduce any resistance from local residents to having a light rail stop as a neighbor, unless it takes away people's front yards. This was a factor which led to the demise of plans in the early 1990s for a light rail system in Edinburgh.

A new light rail system will operate at a high frequency of service, so that waiting times are short and therefore the need for station facilities is limited. Personal security is an issue of perception. The evidence shows that more effective than closed-circuit television is a busy station on a well-used system, where social pressure and passenger cohesion both make travel pleasant and give riders a good feeling of safety.

Figure 3.9 Tram stop in a department store in Ludwigshafen, Germany (Photo: Philipp Krammer)

3.5 Land Use Integration

This is a big issue, whether the local government system is based on a tight control of land use or practices a free market philosophy. Transport is a derived demand from land use activity. There is therefore an argument that coordinating decisions on transport and land use development can make good sense (Figure 3.10). If there is a completely free market and the land use developer and light rail system promoter are the same, for profit maximization reasons, the developer will try to integrate land use development with the introduction of a new light rail system. Indeed, in the 19th century, many cities (e.g., Chicago) were opened up by developers acquiring large tracts of land and then building a rail system, hand in hand with property development.

Some city councils have discretionary powers to "tax" new developments to contribute to transit projects. In the US, some cities and states have "value recapture" taxes on property to recoup part of the rise in land values derived from new (light rail) transit systems or other public investments. In the UK, public authorities can "tax" new developments (e.g., in England permission to build can include an agreement [Section 106] where the developer pays for public benefits) to pay part of the cost of new transit systems that are planned for introduction.

Both of these taxing approaches may be resented by developers, who have little or no say in how their money is spent in the light rail project and may see the project as a public sector toy paid for by the private sector. In

Figure 3.10 Light rail extension and urban development in Amsterdam, 1993 (Photo: A.P.H. Velthoen)

extremis, a developer can look into projects in other cities with more favorable tax regimes.

Local authorities often see transport as a "quick fix" for urban problems. This is because a new transport system can be up and running in less than 10 years, whereas comprehensive land redevelopment may take 50 years, time frames in which governments rarely work. Without care, a new transport system can in fact make an existing land use problem worse, by intensifying development at critical and accessible places and leaving other areas starved of investment, suffering from "planning blight," to become marginal in the urban economy.

Arguing the counter factual case is just as contentious, since without any public intervention (in a new light rail system) parts of the urban economy and fabric will decline and become marginal anyway. In urban development economic theory, this leads to a relative decline in land values until a developer finds the redevelopment of such areas profitable. This was the experience in Liverpool during a 20-year period when the population declined and market prices were lower than new construction.

In all developed countries, local authorities are required to produce development or zoning plans, to guide property developers as to what developments will be acceptable and where they may be undertaken. Such plans may also include proposals for transport system improvements. This can have the effect of guiding developers to particular locations, where transport is likely to be improved, or in some cases to see the transport improvement as part and parcel of the property development.

The impact of these public guidance plans is very critical. On the other hand, a new transport line may have the opposite effect of creating "planning blight," where property owners stop maintaining and developing, pending acquisition and demolition of their property for the new line. When this is a public sector project, implementation can run into decades, worsening "planning blight" to an overall area decline. With the relative decline in land values, land owners receive little or no compensation if property is required for a transit system. This is another source of public discontent.

3.6 Coordination with Property Development

The best way to integrate land use and transport is to have property developers as investors in the new light rail system. This gives the developers a financial incentive both for the investment in transit and to undertake complementary real estate developments. This will enhance the light rail system patronage and hence the total returns to the investors.

It also means that with better access, the land value will also rise. This avoids what is seen as dead money through the taxing approach discussed in Section 3.5 and mobilizes one or more developers. Making this work requires a genuine partnership between the public and private sectors. Although there are many examples in the US and in continental Europe, sadly to date there are few such partnerships in Britain.

Even if there is no direct relationship with property developers, a new light rail line will have an impact on the property market. Research in Germany (Hass-Klau 2004) showed that property prices in the catchment area of a new light rail system rose between 9 and 19%, compared to similar properties further away. In Britain, the picture is more mixed in terms of light rail attracting premium prices for adjacent properties. In Croydon, the additional value created by proximity to light rail was found to be 10%, in Manchester 12%, and in Nottingham 15%.

3.7 Meeting Civic and Environmental Objectives

There is no doubt that a new light rail system can raise significantly the self-image of a city and its perception by outsiders. One measure of this is an analysis of tourist postcards. In cities like Amsterdam, Toronto, and Zurich, light rail is a major feature of the city postcards. London's double-deck red buses are similarly iconic worldwide. Where increasingly retail shops and other establishments are branches of international companies, the unique personality that light rail confers on cities is now an important civic icon and under the control of local people. How important this is to reinforcing cityhood can also be measured by the everyday use made of the services. Even with high car ownership, such light rail systems will be used by virtually everyone, with the usual distribution between frequency of use and percentage of population: a few people use light rail intensely, a few never at all, and most use light rail sometimes.

Most developed countries are now committed to reducing greenhouse gas emissions under the Kyoto Treaty and subsequent agreements. Electrically powered light rail can make an important contribution to city environments. This is especially so if car trips are diverted. If the electricity comes from renewable generation, then the light rail system will reduce both fossil fuel consumption and greenhouse gas emissions.

Light rail emits no air pollution at the point of operation in city streets. It can therefore improve the air quality of the districts through which it passes. This is especially true for the health-threatening emissions: PM_{10} and NO_x. These emissions are in particular connected with respiratory and cardiovascular diseases. In the UK, traffic pollution kills about 45,000 people annually, compared to 3000 killed in road traffic crashes.

Finally, light rail is significantly quieter than internal-combustion-powered modes of transport. Urban noise pollution has been cited as a problem by the World Health Organization. This is both a health problem and degrades the urban social environment. The World Health Organization specifies 55 dBA external noise as the maximum to allow social conversation in city streets. Many busy roads have noise levels greater than 70 dBA, which makes conversation difficult or impossible.

There are many examples of light rail vehicles operating through pedestrianized precincts (Figure 3.11), where other modes of transport would be antisocial. It is difficult to assign a value to the ability of light rail to improve both civic image and prestige. At the same time, it offers real improvements in the quality of life for citizens, both as a means of transport and as a good neighbor. Unfortunately, none of this will be reflected directly

Figure 3.11 Light rail in pedestrian precinct in Nice (Photo: Totally Riviera, www.totallyriviera.com)

in the finances of the system, but it can be an important influence in obtaining the required municipal support and legal approvals. It also can be important in the vital business of attracting inward investment for industry, commerce, or housing, where a city's image can be a major factor in attracting investors (and their spouses).

All the above provides strong justifications for the promoters of light rail to undertake full and genuine public consultation, as discussed in Section 3.1. A city that inherits a new light rail system which has not involved the demolition of property or the disruption of communities during construction will have willing supporters who likely will be regular riders.

3.8 Freight on Light Rail

Historically, many light rail systems carried freight, from parcels on streetcars to dedicated freight cars. Some current light rail systems also carry freight. At the high end, where light rail shares railroad tracks, as in San Diego, regular locomotive-hauled freight trains operate overnight, when the light rail service is suspended. Similarly, many of the old Comecon countries used city streetcar tracks to allow the movement of one or more rail freight wagons, between mainline sidings and factories and distribution depots. Usually small electric locomotives haul the wagons.

More recently, old tramcars have been converted to carry dedicated freight containers in Dresden (Streeter 2009). This CARGO service is an initiative from Volkswagen to link three factories in the city with a merry-go-round service using the streetcar tracks and carrying pallets of parts. This saves an estimated 200 truck movements per day in Dresden. Carrying freight in secure wheeled cages is also practical with low-floor light rail vehicles and level boarding. Such movements would allow local shops to be serviced without the need for trucks parking in the street, reducing traffic congestion and air pollution. Such a freight service would be combined with consolidation centers as proposed by Anderson (2009). Clearly, careful planning of access to any freight depots is required, as is access where wheeled cages are expected to be loaded and unloaded.

4

Engineering Light Rail

Translating the planning stages discussed in Chapter 3 to a project that can be constructed, operated, maintained, and developed requires considerable engineering inputs. The conversion of the planning drawings into engineering drawings is an important and critical step. From engineering drawings, the light rail system can be constructed.

Light rail demands the interaction of three major engineering disciplines: civil to design the structures and tracks, mechanical to design the vehicles, and electrical to design the electrical power system to feed the vehicles and the traction and other electrical systems within the vehicles. Of these, only electrical engineers have a common interface with both civil and mechanical engineers in the design of the infrastructure power and vehicle traction equipment.

Civil engineers have to interface with mechanical engineers regarding the rail/wheel interface, including the resilience of track supports and the suspension of the vehicle body on the wheels. Track quality also impinges on the performance of the pantograph in contacting the overhead line needed to carry the power to the vehicle for traction and auxiliary systems. Electrical and mechanical engineers have to interface regarding the traction motor torque transmission to the bogies and wheels.

These are, of course, not new interfaces, with considerable previous experience and ways to resolve problems for a new light rail system. Incremental developments from earlier projects obviously can be applied where the new system is merely using previous technical solutions. Engineering advances, from new ideas and approaches to both materials used and designs adopted, will require testing and evaluation beforehand. Where such advances can significantly reduce costs, or improve service, then the promoter will have a big incentive to assist in such developments.

Compared to 40 years ago, testing has been made much easier by the use of computer-based analysis tools, for all aspects of the engineering design. Computer-aided design is now a mature area, and packages like AutoCAD and Inventor allow designers to not only undertake detailed designs for their aspects of the project but also to share files with other engineers to consider integrated structural and dynamic interaction between the infrastructure, vehicles, and equipment.

Modeling from satellite measurements and ground-penetrating radar allows the civil engineer to examine many options for the alignment and installation of the light rail tracks. At the same time, the civil engineering team can consider the cost, ease of installation, and minimization of urban disruption during construction.

Mechanical engineers can use finite element analysis to ensure the strength and robustness of body designs and dynamic simulation packages to model the behavior and performance of the vehicles. Electrical engineers can model power consumption and current flows, dependent upon the vehicle design and planning operating practice. In particular, station location and acceleration power needs are dependent on likely passenger loadings and the topography of the line. This allows the design, capacity, and location of substations to be optimized.

These, however, are only applicable where the new designs are an interpolation from earlier experience. Where new designs require an extrapolation of earlier work, then invariably physical testing will also be needed to confirm and calibrate the computer models. Even then, recent experience with light rail vehicle body fatigue problems (Hondius 2007) shows that only physical testing (Figure 4.1) can finally confirm the durability of designs, even when there is an assumed measure of overdesign.

4.1 Design and Other Standards

A new light rail system design will need an overview to ensure that all interfaces are compatible, in particular to avoid creating a problem, the solution to which could be worse than the original problem. Resolving a problem at the design stage should be the aim of the design team, since this is the best time and least cost opportunity.

A well-versed project engineer will lead the team of civil, mechanical, and electrical engineers. The project engineer will identify problems early and then direct resources to avoid or mitigate them. The easiest option for new light rail systems would be to clone a previous system that is clearly working satisfactorily. This would be fine if all cities were identical with exactly the

Figure 4.1 TRAM wheel testing, 1994 (Photo: Ralph Boese)

same set of geographical problems to be considered. Unfortunately, that is never the case, as no two cities are the same.

The "cloning" approach was the one adopted by the new generation of light rail systems in North America opened from 1980, where equipment was obtained nearly off the shelf from Germany. This had the advantage of both extensive system testing and experience and a range of manufacturers to offer proven designs. Especially important was the Frankfurt U2 car, developed in a partnership between the Frankfurt tramway company and the Düwag car company of Dusseldorf. This in particular had a proven monomotor traction bogie on which light rail vehicles (LRVs) are mounted. Nearly all the new systems, including Calgary, Edmonton, San Diego, and Sacramento, adopted this design. Indeed, Siemens, which acquired Düwag, opened an assembly plant in California to supply the North American market.

These first new systems also adopted German designs standards (BöStrab) for tracks, power systems, and signaling and control. By modifying US railroad standards, proven German technology could be used with little or no modification. While the technology may be easy to clone for a new system, more inventive solutions will be needed for other problems, usually network or geographical, which are unique to every city. Without solutions to these geographical problems, a new light rail system cannot be introduced.

The transit industry needs to keep up with developments in other countries, in particular in response to the requirement to provide equal access to passengers with disabilities. The biggest challenge is to accommodate passengers in wheelchairs. The first-generation new American light rail systems had

Figure 4.2 High platform in street of Hannover, Germany (Photo: Urbanrail.net)

adopted high-floor vehicles. Retrofitting these for wheelchair access under the Americans with Disabilities Act was expensive and resulted in the introduction of low-floor vehicles.

The principal reason for the use of low-floor LRVs is that it is difficult and expensive to retrofit accessible stations in city streets for high-floor vehicles, to provide level boarding in central areas (Figure 4.2). This issue was also faced by the Manchester Metrolink, which opted for high-floor vehicles in 1991 and profiled high platforms in city streets to give level boarding at some doors.

The European Union (EU) is working towards the creation of standards for all areas of industry and activity, including light rail. Until that is achieved, there are differing national standards and approaches to the approval of light rail systems. In principle, one country's approved standard should be recognized in all other EU countries. In practice, such interrecognition is far from complete.

An EU-wide set of light rail standards will inevitably be heavily influenced by German practice, since Germany has the largest number of operating tramway and light rail systems and the longest continuous experience. An EU standard also will need to address the difficult task of enabling innovation, without the need to subsequently rewrite the standards. This will mean EU standards based on performance and impact, rather than the more common prescriptive technical standards used in many countries.

4.2 Design Constraints

The principal constraint for design engineers is to take the planning scheme and make it work and be fit for the purpose. Inevitably, compromises are required. In the introduction to this chapter, there was a discussion on the interfaces between engineering disciplines in the design of light rail systems. There are more, and probably more difficult, interfaces between the light rail system and the rest of the city.

The issue of understreet utility plant has already been discussed in Section 2.7. There are also interfaces with highway and traffic engineers over the structure of the roads being used for light rail routes and the way that traffic is managed during system construction. Traffic management after the light rail system has opened and is operating is also important. This is especially true in relation to any reallocation of road space. There are also interfaces between the system electrical engineers and electricity supply companies regarding the required power and the location of substations to feed the light rail network.

Aesthetic and environmental considerations may place a large constraint on the installation and implementation of a light rail system in historic city cores. The intrusion of overhead lines (OHLs) to supply the power to LRVs in sensitive areas has long been an issue for new systems. There are, however, many long-established systems with OHLs, even through sensitive historic areas with World Heritage Site designation status (Figure 4.3).

Figure 4.3 Vienna OHL and Opera House, a World Heritage Site (Photo: Leif Spångberg)

Visual sensitivity to OHLs was one of the reasons for the development in Bordeaux of yet another surface contact system to supply power to the vehicles. As in previous attempts over the last 100 years, it has encountered considerable difficulties. The principle of the Bordeaux "third-rail" system is that a plate in the road is energized when the LRV passes over. This is achieved by a radio signal, and a current-collecting shoe runs over the plate. This costs about 400% more than OHLs. The Bordeaux system was developed by Alstom.

Bombardier has also developed a power supply system that does not need OHLs. Alternating currents are induced in electrical coils set in the roadway between the running rails. Under the LRV is an induction coil, the second part of a transformer using an air gap for the magnetic induction. At the time of this writing, no commercial service is operating with this system.

The architectural dictum of form following function is particularly true for new light rail systems in urban areas. One obvious example is how to support the OHL: by traction poles or building suspension. Establishing the engineering parameters for the OHL to have the right strength and durability is not difficult. If a more costly system, like that adopted in Bordeaux, is used, this also will complicate the design of the tracks.

The main focus should be to ensure that the equipment fits into the urban environment. Two schools of thought prevail. One is a pastiche of 19th century designs: to make the system mimic historic city centers and architecture. The other is to make designs conform with other street furniture (e.g., traffic signs, bus shelters, etc.). Unfortunately, there are examples where neither of the above has been adopted and brutally functional installations were used, like the I beam traction poles in the streets of Croydon (Figure 4.4) and Manchester. These would be more fitting on a private right-of-way or a mainline railway electrification.

What applies to OHLs also applies to track installation, station design, street equipment (e.g., electric junction boxes), and of course the LRVs themselves. Often, vehicle manufacturers will build a mock-up of the proposed LRV and show it off, inviting feedback both from the public but especially from potential riders.

It is also important to get ergonomic feedback from the operators, related primarily to comfort and safety. Thus, it should not be difficult to design LRVs that are comfortable and safe for passengers and the crew. These ergonomic standards, however, need to be interpreted by industrial designers to achieve an aesthetically pleasing finished product.

Earlier, light rail engineers were able to achieve harmonious designs that both were functional and complemented the urban scene. New light rail design engineers have a heavy responsibility not to desecrate city streets and thereby increase public resistance to new systems being opened. Pleasing form that satisfies the function will also win over citizens to become riders.

Figure 4.4 Croydon heavy OHL masts (Photo: L. Lesley)

4.3 Tracks

Without tracks, light rail systems will not work. Tracks confer three important functional advantages:

1. Automatic steering
2. Low rolling resistance, an order of magnitude less than rubber tires on roads
3. Return path for the electric current consumed

In addition, tracks must safely support LRVs, transmitting static and dynamic forces into the ground and also in streets, absorbing the forces imposed by rubber-tire (heavy road) vehicles.

4.3.1 Automatic Steering

Unlike mainline railways, light rail systems have tighter radius curves, often as low as 10 m (30 ft). On railways, wheels steer through the forces created by the conicity of tires and inward inclination of rails. This method of steering, however, only applies to curves with radii above 150 m (500 ft). Many light rail curves, most points, and turnouts have radii about 25 m (80 ft),

Figure 4.5 Worn side rails on Blackpool track curve, 2006 (Photo: L. Lesley)

where vehicles are steered by wheel flanges against the rail side and flange keepers, one of the causes of wheel squeal and rail wear (Figure 4.5). This detail has important impacts on track design and maintenance.

4.3.2 Rolling Resistance

The coefficient of friction (and adhesion) of steel wheels on steel rails lies between 0.1 and 0.3, depending on track, weather, and wheel conditions. Rubber tires on roads have a higher coefficient of friction of about 1.5, added to which is the fact that pneumatic tires flex as they roll, absorbing further energy, raising the temperature of the wheel. The low rolling resistance of LRVs means that most of the power consumed is for accelerating the vehicle mass and can be determined from:

$$KE = 0.5\ mv^2 \tag{4.1}$$

where KE = kinetic energy (Joules), m = mass (kilograms), and v = speed (meters per second).

This is an important design parameter for a new light rail system, to help reduce power consumption. One way to achieve this is by lighter vehicles. This will reduce acceleration energy and also that expended in rolling resistance. As a comparison, a double-deck bus with a passenger capacity of 100 weighs about 12 tonnes empty. A typical LRV with a capacity of 200 weighs about 40 tonnes empty.

If they had the same weight per passenger, the LRV would weigh 24 tonnes. Ironically, buses use about the same energy per passenger as light rail because the lower rolling resistance of LRVs is lost by the greater kinetic energy needed to accelerate the heavier LRV. Diesel buses, of course, need oil, while LRVs use electricity, which can be generated from a variety of fuels and from renewable sources.

4.3.3 Return Current

In all light rail systems today, the tracks are part of the electrical circuit between LRV and substation. This was not always the case. Until 1962, the Washington, D.C. system used an underground current supply system known as the "conduit." It had a slot between the running rails, through which a pickup passed. The rails had no function in the traction current circuit.

In comparison to the total electrical resistance of the traction circuit between substation and LRV, the rails are at least an order of magnitude lower than the OHL. This means that, provided the rails are fully bonded electrically, there will be few problems from stray currents. This is discussed more fully in Section 4.6.

4.3.4 Embedded Tracks

There are two classes of light rail tracks:

1. Embedded in highway pavements and shared with road traffic
2. Segregated and for exclusive use of rail vehicles

Light rail tracks in streets are normally laid with the railhead horizontal and the rail web vertical. This minimizes the impact of rubber-tire road vehicles on the rails. Heavy road vehicles are normally more damaging than LRVs. Tracks embedded in highways need to be able to transmit the static and dynamic forces from the railhead into the subbase.

As traditional grooved girder rails are about 7 in. (180 mm) high, this usually means a foundation about 12 to 20 in. (300 to 500 mm) deep below the rails for support. The rails also need to be kept to gauge, which for most of the history of tramways has been achieved by bars bolted through the web of the rails, allowing for lateral adjustment.

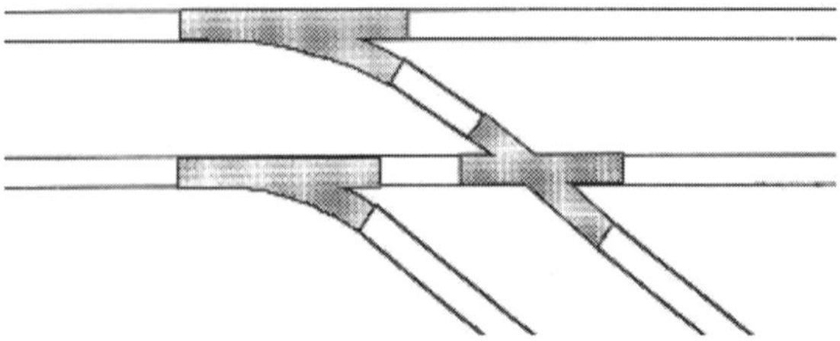

Figure 4.6 Turnout units (Drawing: L. Lesley)

Switches and crossings, or turnouts, will normally be a special design for the particular street location and junctions, to fit in with the street corner curvature and to protect the footway at the side of the carriageway. These are normally prefabricated in a rail factory, with three main units (Figure 4.6), two of which are the same (switch unit). Plain rails will join the three units to make a complete turnout for a single track. For a double-track turnout, ten units (four switches and six crossings) are needed. Prior to shipping to the site, the complete fabrication will be assembled in the factory, to ensure that it matches the detailed track drawings and thus can be reassembled at the (road) junction correctly.

The highway provides the foundation for streetcar tracks (Figure 4.7). For this to be successful, the pavement and its subbase must be capable of carrying the static and dynamic loadings of LRVs. Pavement design considerations are well understood and rest upon the suitability of the ground under the road. The universally accepted measure of ground strength is the Cali-

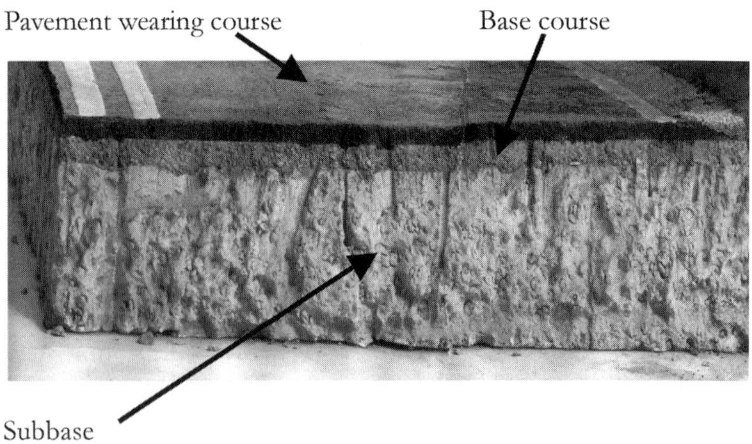

Figure 4.7 Typical section of a flexible highway pavement (Photo: L. Lesley)

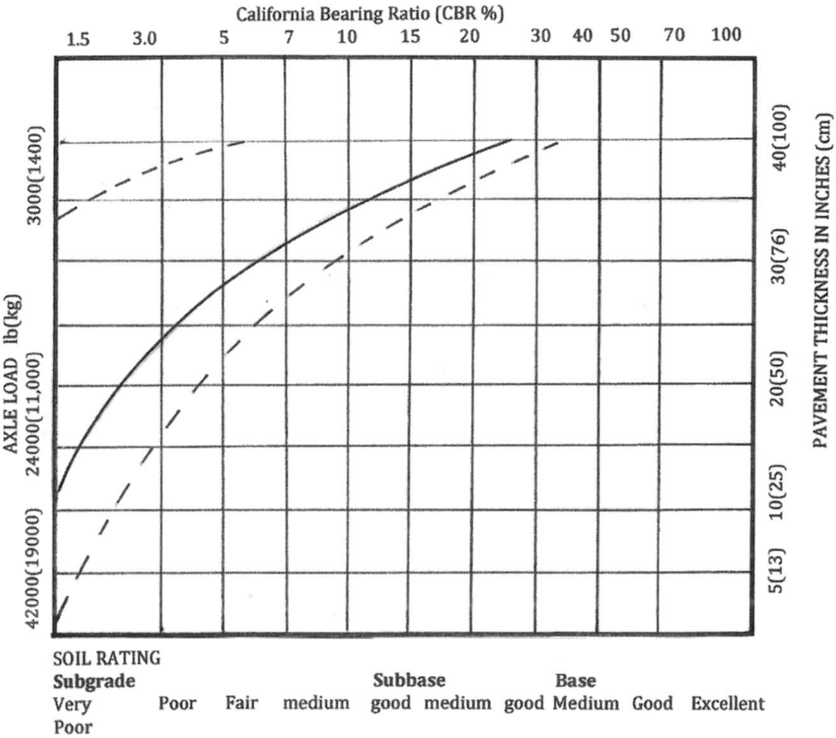

Figure 4.8 CBR, pavement thickness, and traffic (Drawing: L. Lesley)

fornia bearing ratio (CBR). The lower the CBR, the thicker the pavement has to be for a given volume of traffic, with maximum axle weight and expected total life (Atkins 1980) (Figure 4.8).

The pavement thickness of a downtown or arterial road carrying 30,000 or more vehicles per day will be between 12 and 15 in. (300 to 380 mm) (Figure 4.8), depending on the CBR. Provided there are no surface signs of pavement failure, such as rutting (fatigue), potholes (subsidence), or wearing course needs replacing, then the subbase will have a CBR more than adequate to support streetcar or tramway tracks, whether using the orthodox method, with a monolithic concrete slab, or newer less intrusive designs, where the strength of the road base is used as the foundation for the track. This consists of a separate prefabricated concrete longitudinal beam for each rail, only 7 in. (180 mm) deep and 14 in. (360 mm) wide (Lesley 2002) (see Figure 4.10).

Both methods require the road pavement to be excavated to allow tracks to be laid. The slab track (orthodox) method is more intrusive, takes longer, and requires the understreet utilities to be removed from under the new track slab and then relocated. The design and coordination of utility diversions are almost worthy of a text in their own right. Thankfully, the utility companies are able to do that but need to be paid and scheduled into the track installation program.

Closing streets to make this possible requires careful traffic management, as well as steps to maintain access required to premises along the route. The approach adopted (e.g., a block-by-block installation or the whole downtown as one installation) will determine how comprehensive and long-lasting the disruption has to be.

Ideally, the traffic rerouting needed for track installation should be the same as the ultimate traffic management regime when the light rail system is operating. This is both easier to achieve and will make it simpler for the ultimate traffic priorities afforded to the light rail system to be adopted and accepted by motor car drivers.

Once cleared, excavating the road is a major undertaking, since even if some of the spoil can be reused on-site, it has to be stored to create a working space. Clearing away the road pavement can be undertaken with backhoes or similar excavators. Excavating the utilities, unless they have already been decommissioned, is nearly a hand excavation operation. If the utilities have been decommissioned, then any plant below the foundation slab level can be left *in situ*.

Backfilling and compacting the road base in readiness for casting the concrete slab has to be done in stages, with spoil laid in 3-in. (75-mm) layers and compacted to a CBR of at least 5%. A track (support slab) is cast on this *in situ*. This can be done with temporary shuttering in sections. Alternately, it can be slipform paved with a one-pass machine. This can include, as in Sheffield, continuous slots to take the rails.

When the concrete is cured, the rails are welded into long strings, for electrical continuity, and then clipped to the slab. Finally, the pavement running surface is replaced up to the level of the top of the rails (Figure 4.9).

The success of this method relies on the rails being rigidly fixed to the concrete foundation slab, usually by rag bolts or similar clips, and the surfacing forming a watertight seal to the rails, to prevent water penetrating into the pavement. In practice, over time the fixings work loose, and heavy road vehicles rock the rails and break the seal with the pavement. Once broken, any surface water is then pumped into the pavement by passing heavy-vehicle tires, further damaging the installation and ultimately leading to the failure of the pavement or the rail fixing or, in the worse case, both.

A failed pavement will not only give an uncomfortable ride for road vehicles but with potholing be potentially dangerous. There will be special arrangements in different cities as to who is responsible for the maintenance of the highway pavement around light rail tracks. The light rail operator will,

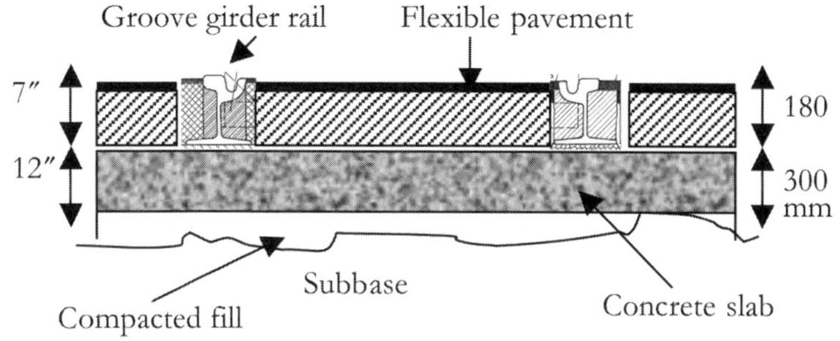

Figure 4.9 Orthodox street-embedded tracks (Drawing: L. Lesley)

however, always be responsible for maintaining the tracks in a safe condition. From experience across Europe's 300 light rail systems, it is heavy rail vehicles that are the principal cause of damage to embedded street tracks.

For this reason, a new embedded track system was developed to exploit the existing road pavement, rather than having to rebuild the highway structure, and also to avoid the need to relocate understreet utilities. The LR55 (Figure 4.10) system separates the structural functions of a girder rail. LR55 rails use the vertical strength to supporting concrete beams, which are very stiff vertically and horizontally. The LR55 rail retains the LRV steel wheel support and guidance functions.

This results in a shallower and lighter rail and overall a much less intrusive installation. The key to making this work is the use of a polymer to bond the rail into the concrete beam. The polymer will be strong enough to resist thermal forces in the rail. It also provides an elastic support structure for the

Schematic

Transition to street track in Sheffield

Figure 4.10 LR55 strip beam foundation tracks (Schematic: D. White; photo: L. Lesley)

rail. This has the benefit of reducing railhead damage, like corrugations, to almost nil and gives a smoother ride for rail vehicles and their passengers.

The LR55 system is also more robust for heavy road vehicles like buses, as it is both more resistant to the kind of damage that Phoenix rails (e.g., Ri60; see Figure 4.33) with thin keepers suffer and also does not jolt vehicles crossing or going along the tracks.

The mass-spring-mass-spring arrangement reduces airborne noise (at 100 Hz by about 20 dBA) and ground-borne vibrations (at <20 Hz by about 30 dBA). The LR55 track installation is also simpler and quicker, as it requires only a minimum of pavement excavation. Using prefabricated concrete beams means that a section of road needs to be closed for about a week while a minimum of 100 m (110 yd) of track is laid. With practice, a track gang could lay up to 100 m per 8-hour shift.

As part of the testing program for the LR55 track system, it was installed in the entrance of the Rotherham Bus Station (Figure 4.11), with 2500 buses per day. During its 2-year test period, more than 2 million buses passed over, along, and scrubbed across the rail. This was the equivalent of the bus traffic on a busy European urban road over a 30-year period. From this, the LR55 track was used to replace a failed section of 80-lb rails on the Sheffield Supertramway.

The rounded head of the LR55 rail also means that heavy road vehicles crossing the tracks do not face a sharp upstand of the groove keeper. In the UK, this has led to the rails being damaged, potentially exposing LRVs to the danger of derailment (see Figures 4.12 and 4.47). The Sheffield installation of the LR55 track replaced an orthodox rail design using 80-lb rails that had failed in less than a year after opening in 1994. The LR55 was installed over a weekend in March 1996 and has remained without any maintenance since. Further, with the continuous elastic support, LRVs on the LR55 section of track are quieter and no corrugations have formed (Lesley 2009a). Also, the railhead shows little sign of wear, and the rail should last a minimum of 30 or probably 40 years without attention. Finally, the polyurethane bonding is not delaminating despite 100 heavy road vehicles (44 tonnes gross vehicle weight) crossing daily.

The LR55 track system has been comprehensively tested (Lesley 2002). This has included cyclic testing at main line axle loadings (25 tonnes 200 million), up to 80-tonne axle loads, failure tests over voids (60 tonnes), and pullout (30 tonnes). Testing was conducted underwater and in temperatures between –10°C and +60°C. Electrical resistivity testing was undertaken in laboratory conditions and on field samples. Live testing was undertaken on the Rotherham Bus Station, impacted by 2.5 million buses, the equivalent of at least 30 years on a main city street. A length of 80-lb railed track of the Sheffield Supertramway failed after 1 year's service and was replaced by LR55 in March 1996. Since then it has provided maintenance-free service,

Figure 4.11 Rotherham Bus Station LR55 installation (Photos: L. Lesley)

with 3 million tonnes of LRVs and 1.25 million tonnes of heavy road vehicles passing over annually.

In the absence of a different approach to orthodox embedded track design and installation, damage from heavy road vehicles (Figure 4.12) will be a permanent, disruptive, and expensive maintenance cost. This is in addition to the higher initial cost of installation An example is the 3.4-km (2-mile) street tracks in Manchester, installed in 1991 over 52 weeks, which cost £8 million (US$13 million) to relocate understreet utilities and £6 million (US$10 million) for the track installation. In addition to regular patch and repair of these tracks, the entire street track network in central Manchester was replaced during the summer of 2009, at an undisclosed cost. The estimate for laying the Manchester city center tracks with LR55 was £7 million and using only one track gang (not three) would have taken just 34 weeks.

Figure 4.12 Track damage in Manchester, 2009 (Photo: L. Lesley)

4.3.5 Segregated Rights-of-Way

The most economic construction for light rail on segregated rights-of-way is a classical rail track. It has rails clipped to ties (sleepers) on crushed rock ballast (Figure 4.13). Some early pioneers of segregated light rail rights-of-way (e.g., Winnipeg and Liverpool in 1914) tried to soften the visual impact of railway tracks by turfing with grass. Unfortunately, the grass retained rainwater, which rotted the wooden sleepers used. More recently, German designs using concrete units to support the rails allow grass without shortening the track life or increasing maintenance costs (see Figure 4.15).

Segregated rights-of-way are often in the middle of dual carriageway roads but can also be found alongside roads and completely away from roads. Older or abandoned rights-of-way, canals (e.g., New Jersey), or railways (e.g., Manchester, San Diego) can be used. The cost of creating new rights-of-way in dense urban areas, including buying property, relocating residents and businesses, and demolition, is very high and therefore is best avoided if possible.

Although segregated light rail rights-of-way using the medians of divided highways was pioneered in North America and England in 1914 (Horne and Maund 1982), it was taken up on a citywide basis in Liverpool. The visionary city engineer John Brodie planned the expansion of the city on the basis of 120-ft- (37-m-) wide radial and ring roads. These new roads were all built

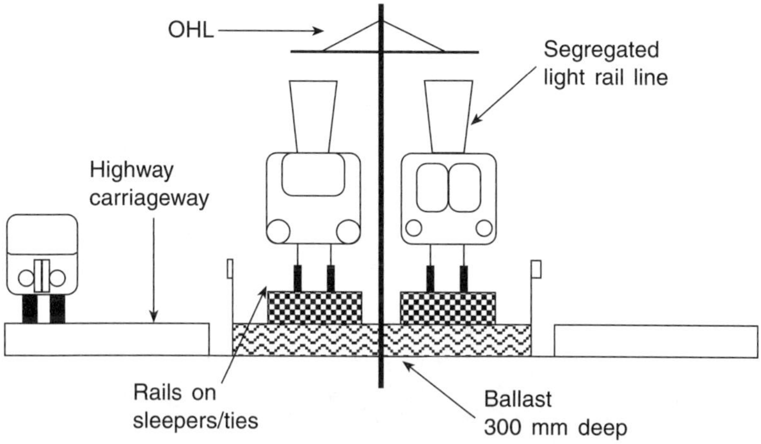

Figure 4.13 Schematic of construction of segregated tracks (Drawing: L. Lesley)

with a grassed median strip 33 ft (10 m) wide. The first light rail line was opened in 1914, using a clinker ballast base to support wooden sleepers (ties) 9 ft × 10 in. × 6 in. (2.74 m × 0.25 m × 0.15 m) and infilled with clinker. Soil with grass seed sown was laid on top to just under the railhead. Sadly, there was inadequate drainage, which led to the sleepers rotting prematurely.

Nevertheless, the benefits of segregated tracks led to this work being taken up in earnest after the Second World War, as cities in Europe rebuilt from war damage. During this reconstruction, tracks were relaid separated from the road carriageway in a reserved section of the road (reservation). This led to the improvement of the economics and service of the tramways and was especially so for low-cost extensions into new residential areas on the suburban fringe. During this period, improvements to the engineering of segregated tracks followed.

The principal engineering improvements involved better prepared track beds, graded to ensure that rainwater drained away. Where the ground cannot absorb the water, special drains are built, with the use of larger sized (more than 1 in./25 mm) aggregate for the ballast, to improve drainage. Concrete sleepers now nearly predominate, although in the spring of 2009 wooden sleepers were still being used in street track renewals in Budapest (see Figure 4.32).

Finally, for segregated tracks, conventional (grooved or ungrooved) rails are clipped either directly or via a base plate to the sleeper, with the rails long welded to give a smooth ride and electrical continuity. In this case, because the rails are exposed to the weather, precautions have to be taken to lay them at a neutral temperature or to compensate by tensioning them to prevent buckling in hot weather.

Figure 4.14 Road lane reservation in Szczecin, Poland (Photo: Neil Pulling)

Where new roads are being constructed, a median strip wide enough for light rail can be provided at a marginal extra cost. In the near future, few major urban dual carriageway roads are likely to be built, so segregation of light rail tracks will have to exploit existing roads. This can be achieved by reallocating carriageway space on the basis of the capacity provided (Figure 4.14). A typical light rail system operating on a 5-minute frequency with 200-passenger-capacity vehicles will have a peak capacity of 2400 passengers per lane per hour.

The same space on a freeway will have a vehicle capacity of 1500 cars per hour, at a peak period average occupancy of 1.2 passengers per car, which means an hourly flow of 1800 people. On a radial city street, a peak flow of about 800 cars per hour can be expected, giving a passenger capacity of under 1000 per hour. Thus, light rail will have a 33% capacity increase compared to a multilane freeway and be more than 50% better than a four-lane city street. Converting one of two road lanes to light rail improves the peak passenger capacity from 900 to 2850 per hour, a 300% gain.

This approach was pioneered in Amsterdam, where other traffic was removed from light rail tracks, and the tracks were raised about 4 in. (100 mm) above the carriageway, with a chamfered curb to denote the difference and exclusivity. Such "trambaans" also have another positive benefit: they provide congestion-free routes for emergency services (police, fire, and ambulance) to reach incidents quickly, saving time and lives. Figure 4.14 shows an example from Poland where only white lines 150 mm (6 in.) wide separate LRVs from general traffic on an inner city street.

Provided the light rail service is frequent, such surface reservations are normally self-enforcing, without any legal penalties. In the peak period, the sight of LRVs passing stationary lines of congested cars is usually a good advertisement to frustrated car drivers. At least at the margins, enough will switch to light rail to reduce the congestion. Having convenient park-and-ride is normally the key to making this switch successful.

New light rail systems built into the suburban parts of a city can use rights-of-way separate from highway traffic but close enough to residential areas to make the service convenient for residents living along nearby roads. Figure 4.15 shows an example of a new line in Madrid, where the light rail line is alongside a highway. The track is grassed, so that most of the time when there is not an LRV passing, it looks like a park and helps to absorb some of the carbon dioxide and other fumes from the adjacent road traffic.

The light rail engineer can now demonstrate how a new urban transport facility can be a friend of the earth and also a friend of the system funder, by economical designs. Infrastructure that has a long life and low maintenance cost will be the cornerstone of a viable system. More importantly, by building a convenient and comfortable line with high-frequency and reliable service,

Figure 4.15 Lawn light rail tracks in Madrid (Photo: UITP)

the engineer will maximize ridership, including attraction from and reduction of motor car traffic.

4.3.6 Rail Problems

Different combination and methods of track construction will be used in most light rail systems to accommodate different rights-of-way. The action of steel wheels running over steel rails creates wear and failure problems, of which the track engineer must be aware. In ideal conditions with regular maintenance, rails on straight tracks can last 40 years. Curved tracks create the most problems for rail life and safe operation. LRVs will tend to grind the vertical face of the (inner) rails and in extremis (see Figure 4.6) can not only put the track gauge out of tolerance but also create conditions where wheels can climb the rail face onto the rail top and derail.

Track or wheel lubrication can help to minimize this rail side wear, but not eliminate it. This requires the regular reprofiling of the rail, which could, however, bring the track out of gauge tolerance. In this case, swapping the inside and outside rails to use the unworn rail sides will extend the life of the track. Ultimately, rails will have to be replaced on severe curves typically between every 8 and 10 years.

The most likely problems for the track engineer relate to fatigue. The most common of these is the creation of corrugations on the railhead (Lesley 2009a). These are caused by unstable running along the track, creating work-hardened zones at a regular interval (Figure 4.16). While this is not usually a safety-critical failure, it generates uncomfortable riding inside the vehicle and can damage the vehicle suspension. Corrugations also generate noise, which is a considerable nuisance to those living or working alongside the tracks.

Eliminating the cause of corrugation formation requires the wheel to be kept exactly to profile in order to reduce unstable running. Once formed, only railhead grinding is possible, which has to remove about 0.008 in. (2 mm) to prevent the roots of the corrugations from reaching the bottom of the railhead. Unless this is done regularly, corrugations can only be eliminated by rail replacement. Regular grinding of the railhead also helps to keep a profile that results in more stable running and therefore is less likely to lead to the formation of corrugations. This remedial action, however, shortens the rail life.

Another area to which the design engineer will need to pay attention is the turnouts (switches and crossings). This will be particularly so if there is uneven light rail traffic between the branches over a crossing. The nose of the crossing will be subject to differential wear and damage. The nature of the damage will be determined by whether a single casting has been used (Figure 4.17) or the crossing has been fabricated (see Figure 4.48). Fabricated

Figure 4.16 Corrugated track (Photo: L. Lesley)

Figure 4.17 Crossing in Sheffield (Photo: L. Lesley)

crossings are also more vulnerable to damage from heavy road vehicles. This can be minimized in the track alignment planning by avoiding the need for heavy road vehicles to pass over a crossing. Damage to street track crossings will also be affected by whether the groove is raised through the crossing and LRV wheels pass over on their flanges. If wheels pass over on their tires, the nose of the crossing will be impacted and subject to damage by wheels. This is unlikely where wheels pass over a crossing on their flanges. Compare the wear of a cast crossing in Sheffield (Figure 4.17) to a similar aged fabricated crossing in Sacramento (see Figure 4.48).

4.4 Stops and Stations

Stops and stations perform two important functions. First, they are the interface between passengers who want to travel and the vehicles that provide the travel service. The second function is to provide a timetable observance point, so that the system can operate reliably and regularly. The crucial planning and engineering question is how many stations and where they should be located.

The academic and engineering literature includes many papers on station optimization. Most, however, are based on uniform density catchment areas (e.g., Lesley 1976). Few cities have uniform density catchment areas, so these papers can only be a guide to actual station spacing. In practice, urban areas are not uniform, so identifying station sites will require considerable local knowledge and research.

The other approach is to assess the market. How far are people prepared to walk to reach a light rail or tramway stop? Here the literature is clearer cut: 500 m and often up to 800 m. In a catchment area with an 800-m radius, the average walking distance to the station will be about 530 m. When generalized cost assessments of walking are considered, then the light rail planner is seeking to minimize walking distance, since this is perceived to be twice as irksome as time riding in the vehicle.

Here one of many design conflicts emerges. To make walking distances shorter, stops should be close together. When stops are close together, the average LRV service speed is reduced, lengthening the in-vehicle time and making the overall trip less attractive. Further, the slower the operating speed, the more vehicles that are needed to provide a given service frequency, increasing the need for staff and therefore overall costs.

Generalized costs and market research exercises show that waiting at stops is even more irksome than walking to and from stops. For reliably regular service, the average waiting time can be shown to be half the service frequency interval. Therefore, the operator has a large cost and patronage incentive to achieve a high service speed, both to maximize vehicle productivity and to generate or attract the maximum patronage and hence revenue.

When putting all these factors together, the light rail system engineer and planner should consider stop spacings of about 500 m, which gives a short walking catchment area and can offer high operating speeds of about 25 km/hour. The key, however, is a careful assessment of local conditions. In suburban and low-density residential areas, longer stop spacings may be both possible and desirable. This is especially true if park-and-ride (both car and bike) is considered, as this maintains short access times to light rail stations that are farther apart, increasing service speed and hence reducing journey times.

In the more densely developed inner and central parts of a city, stops can be as close as 250 m, especially if the route alignment includes curves, where vehicles must slow down anyway. The shorter stop spacing will not significantly reduce the operating speed on that part of the line. It will let the system offer stops more convenient to important traffic generators, like shopping centers, universities, hospitals, cinemas, office complexes, etc., than would a subway service. Such a system will certainly have better accessibility and therefore generate more patronage, as proven by many light rail systems.

The final point on station spacing and location relates to the network attributes, especially if crossing busy highway intersections, where stops may be required for safety and other reasons. In terms of accessibility, highway intersections are the most connected parts of the network, the "central place" in Central Place Theory (Christaller 1933), where historically high-value establishments like banks, department stores, etc. locate.

For the light rail system designer, these are the most valuable positions for stops, as they give the maximum accessibility for the local catchment area. Exactly how a stop should be designed and located will depend on geographical and topological conditions and what, if any, traffic priority the light rail system enjoys over other traffic (Figure 4.18). Ideally, an approaching LRV should have a preempted traffic signal and a clear run into or out of the station, so that only a stop to deliver or pick up passengers is needed. The maximum service speed can therefore be maintained.

After vehicles, the stops and stations are the most public face of a light rail system. Ideally, passengers should only have a minimum wait, at the most a couple of minutes, for the next departure. This wait should also be in comfort and safety, with reliable information. Passengers need both reassurance as well as guidance on how best to use the light rail system. This is particularly true of new systems or where new lines are built to attract car commuters who have little or no previous experience with public transport system use. With frequent service, waiting times will be short, so elaborate waiting rooms with seats are not needed. In cities with high rainfall, shelter

Figure 4.18 Light rail stop in central Nottingham (Photo: L. Lesley)

for a peak accumulation of passengers will be needed (Figure 4.18). In cities with hot weather and strong sun, shade will also be important for the comfort of passengers and may be provided by trees.

As the most permanent and visible part of a light rail system, stops and stations need to be good neighbors. Local residents will not appreciate loud public address announcements late at night. The design of stations is both part of the corporate image of the system and functionally important to attract and retain riders. Considerable design effort should therefore be expended to build stations that enhance the local neighborhood, which would make them more likely to be adopted by residents as part of their civic infrastructure.

In addition to attracting passengers to use the light rail system, stations are also the places where passengers alight to reach their destinations. Here, navigation from the stop must be clear and simple, especially if it includes interchanges with other modes of transport (e.g., buses and taxis). Some stations will also have park-and-ride, so passengers must be able to find their vehicles easily and with certainty. Passengers must be able to leave their vehicles with complete confidence that they will be there on return and will be undamaged.

Design of the parking area must use the principles of defensible space and not just rely on security systems like closed-circuit television, which has a dubious record of preventing damage or theft and does not always provide images that can be accepted as legally valid in court proceedings if wrongdoers are caught. Defensible space includes real people, whether passengers passing through, operating staff on duty, or neighbors who have a clear view of the area. Any person with criminal intentions must be made aware that people will be watching, since being seen or caught by a real person is the best deterrent against crime.

Security and passenger service at each station used to be achieved by on-duty light rail staff. Very few light rail systems can afford dedicated operational station staff. This is not to say that people cannot be employed at stations, especially where heavy passenger flows are expected. Franchises can be offered for ancillary services like newspaper/magazine sales, food and beverage, etc. This would be a way for passengers to see that other people are around and also can help to form part of the reassuring relationship that is needed to turn car commuters into transit users and to make a success of a light rail system.

A light rail station franchisee might also perform some minor system functions, on a contractual basis, at much less cost than a dedicated staff member at the station or a roving staff team, as well as serve as an alert system for security if there is a problem at the station or call emergency services should there be a passenger accident. Franchisees can also sell light rail tickets and even keep the platform area clean.

The above discussion focused on the functional needs of stations to attract and satisfy passengers. There are more prosaic needs in designing and building stations. These revolve around the kind of vehicles that the system will use, the likely peak volume of passengers, and how they will reach the station. For new light rail systems, it is likely that vehicles will be partly or completely low floor, with level boarding from platforms about 12 in. (300 mm) high rather than metro (subway) platforms 3 ft (900 mm) high.

To make it easy for all passengers to board and alight safely, platforms need to be nearly level with the doorway thresholds. There has to be some tolerance here, since as vehicle wheels and rails wear, the relative position of the floor/door threshold will change. Some LRVs have adjustable suspensions that can accommodate wear.

Nevertheless, it is better to allow some tolerance. Typically, as much as 2 in. (50 mm) has been found from ergonomic studies as the maximum safe vertical difference between platform and car floor that elderly people can safely manage. With a platform 10 in. (250 mm) high, cars with floors 300 mm from the top of the rail can operate safely. This is also the maximum tolerance under various disabled access regulations, with typically a maximum 2-in. (50-mm) horizontal tolerance (Figure 4.19).

Platform length will depend on vehicle length and whether at some stage trains of LRVs will be operated. The most prudent approach would be to build to the required system specification on opening. This will accommodate

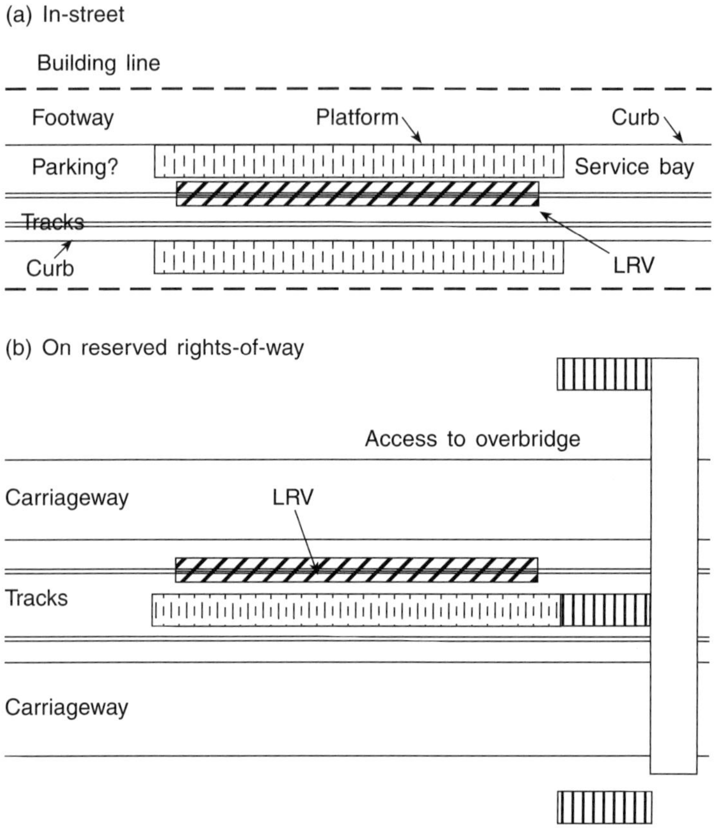

Figure 4.19 Light rail station considerations: (a) in-street and (b) on reserved rights-of-way (Drawing: L. Lesley)

the longest car to be operated. Space can be reserved for future expansion, required by either increased passenger numbers from the initial system or the addition of new lines sharing tracks and stations.

The minimum platform width is normally dependent on regulatory requirements. A new system should aim to provide wide platforms, to give a sense of openness and safety. Typically, a free-standing platform will be at least 6 ft (~2000 mm) wide and ideally 10 ft (3000 mm). It is better if the station can be integrated into the streetscape using footways or sidewalks (Figure 4.18). The formal demarcation between the platform and the functional space available from the footway would allow passenger buildups without overcrowding. This will also be the most convenient and safest way to access the station platform.

Almost by definition, on average 50% of the passengers will originate on the opposite side of the station from which they depart, and vice versa, and therefore will need to cross the tracks to reach their destination after alighting. Historically, heavy railways have made passenger crossings of tracks grade separated, by having bridges or subways, as at Virginia on the San Jose system. Here, however, by using an island platform, a spacious waiting environment is provided, and interchange at an island platform is also convenient. Being in the median of a busy freeway means that an overbridge is still needed for a grade separation and safe access to the station from its catchment area.

Unless there is some particular local problem, light rail systems normally allow passengers to cross the tracks at grade. The most obvious reasons for this are that it is preferred by passengers and is easiest for those in wheelchairs or pushing baby carriages. As LRVs stop at most if not all stations, with powerful emergency brakes, there is unlikely to be many safety issues.

Passengers using a light rail service are more likely to be involved in a safety incident with other road vehicles while gaining access to the station. Having surface crossings of tracks also reduces the size of the platform needed, since space is not taken up with stairs/ramps, up to a bridge or down to a subway. Finally, considerable urban security research shows that most people do not like subways, due to the claustrophobic feeling or perceived lack of security. Pedestrian subways or bridges are also difficult to make accessible to passengers in wheelchairs.

About 80% of the adult population can make a step of 250 mm (above the track) to reach the platform and 90% can make 200-mm steps. Station design should, however, guide the passengers to cross at dedicated crossing points, normally at one or both ends of the platform. Depending on location, the tracks may be paved or open and ballasted. In either case, designs should discourage passengers from walking along the tracks. This usually can be achieved by placing a fence between the tracks at stations where the station tracks are not shared with other road vehicles.

Where interchange with buses is envisaged, ideally it should be cross-platform in the peak flow direction, since interchange is always a perceived penalty in all market research and hence generalized cost calculations. Making interchange easy for passengers without imposing difficulties on the operating staff is the trick the wise design engineer will be able to perform.

Finally, platforms should be straight so that the operating staff in the driving cab can fully supervise the safe alighting from and boarding of the LRVs. Minor redesign of curved approach tracks and possibly tapering the platform ends may be required. This will accommodate the LRV swept path or dynamic kinetic envelope on entering and leaving the station. Straight platforms with tapered ends are, however, better than curved platforms, where it will be difficult to maintain safe gaps for passengers, including those in wheelchairs, to board and alight.

4.5 Depot

LRVs must be stabled overnight, maintained, and repaired. This requires depot facilities. Here the light rail system planner is under a very severe constraint compared to a bus service planner. Bus depots can be located almost anywhere in an urban area and still be able to service a number of routes. The tram or light rail depot must be physically connected by track to the passenger service operating lines. Operating between the line and the depot is a dead cost and generates no revenue. Depot access also takes time out of the driving shift.

In designing a light rail line, the engineer and planner need to be able to find suitable and low-value sites where a depot can be located close to or preferably next to the line. At the same time, because a depot is semi-industrial, it should not be in a residential area, where it will be seen as a poor neighbor.

In some cities, brownfield or derelict industrial sites are available, and locating a new depot there could be used as a fulcrum to help stimulate urban regeneration. Indeed, a deal might be struck with a property developer to use the ground level of a site for the depot and the air rights for commercial and other development. This would have the benefit of excellent access to a new light rail line.

Where such sites are not obviously available, the depot may end up dictating part of the alignment of the light rail line. It would make considerable economic sense to divert the line to be closer to the depot. There would be a trade-off between a reduction in (depot) capital costs and dead mileage in running between the depot and the system and a different pattern of patronage. The diverted line may be not the most preferred route, but if it does not have a suitable depot site, the ideal route would only be "academic." The preferred route can be opened later, linked via the city center to the first route with the depot.

For new systems, the depot site should be chosen to allow for subsequent network expansion, since this will bring economies of scale in terms of technical staff, stores, and maintenance equipment (Figure 4.20). Space to lay additional overnight storage tracks is a low-cost investment that later may have a very high value because it is not necessary to build another depot on a new site.

Once an initial line has been built and opened, subsequent network extensions might include links that were omitted from the first line, in order to have an economical depot site. This almost certainly will include some

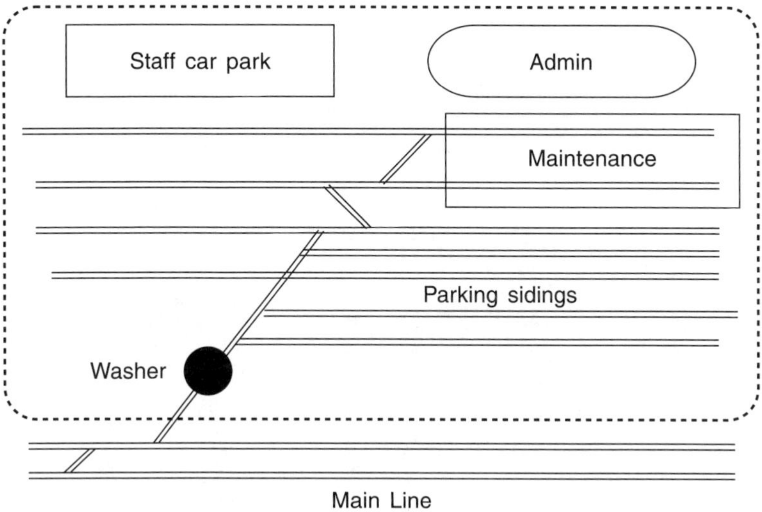

Figure 4.20 Typical depot layout (Drawing: L. Lesley)

shared trackage and an increase in service frequency on a trunk section of line, which will increase patronage and revenue.

Often, new systems are overindulged in terms of depot facilities. Clearly, there are basic daily, weekly, monthly, and annual maintenance activities. There also may be a contract with the vehicle supplier to undertake heavy maintenance elsewhere. Daily maintenance will include washing the outside of the vehicles and cleaning the inside. Normally, a drive-through washing machine is used. It can be external or internal, although an internal washer may be preferred for environmental reasons (Figure 4.21). There are many vehicle washers on the market, some of which are proprietary. The location of the washing facility will be determined by the operational procedure in the depot.

Many systems wash their vehicles at the end of the day's schedule as vehicles enter the depot. Similarly, most systems clean the inside of vehicles using overnight cleaning teams, so that vehicles are ready for passenger service first thing in the morning. Other systems wash the LRVs as they leave for service.

For those systems which have vehicle sand applicators, the checking and refilling of sand boxes normally will have to be located on the inbound tracks prior to overnight parking. Again, proprietary equipment is available to keep sand dry in the depot and deliver it by pipeline under air pressure directly into the LRV sand box.

During the operating day, drivers will report minor faults and problems with their vehicles. The depot maintenance manager will assess the need to

Figure 4.21 Depot vehicle washer (Photo: Tepo-Auto Ltd., Beijing)

examine specific vehicles on return. Obviously, vehicles that fail and have been towed back to the depot will be allocated workshop space for repair or rectification. Other more minor problems may be examined by a fitter or mechanic on the inbound track and either repaired there or diagnosed, with remedial action determined to be undertaken in the workshop on a rotation basis.

These functions all require space and facilities to be performed safely and efficiently. Workshop space under cover and heated/cooled for repair and maintenance will both make for a better workforce and improve the quality and reliability of maintenance and hence fleet operations and performance. The key dimension for this workshop space is the size of the vehicles to be maintained and whether they easily can be split into smaller parts. Most LRVs today are at least 30 m long, so allowing for space at either end means a depot length of a minimum of 40 m. The second dimension is workshop width. This will be determined by fleet size and therefore the planned spare vehicle and maintenance program. Normally, however, such LRVs are in several sections and may be splitable, thus reducing the length of the dedicated workshop section or enabling maintenance to be performed on more than one vehicle on each track at the same time. This is especially so when working under vehicles.

Workshop pits are the conventional way to undertake repairs and maintenance beneath LRVs (Figure 4.22). Pits, however, have their drawbacks,

Figure 4.22 Blackpool depot pit area (Photo: L. Lesley)

including safety and inflexibility. They also can flood and will need regular and expensive cleaning to avoid the buildup of oil and other dangerous debris. If pits are deep enough to walk along, then moveable staging will be needed in order to be able to work comfortably and safely under LRVs. If, on the other hand, the pits are designed for a comfortable working height, then walking along them will be difficult with an LRV on the tracks, since most systems will not allow staff to be in pits under LRVs that are moving.

Another approach to depots is a pitless workshop, with vehicle jacks. These allow vehicles to be lifted to a safe working height off the tracks and can be adjusted depending on the task to be undertaken. This is particularly valuable when bogies need to be removed or replaced. A flat depot floor with jacks to lift vehicles will require less shunting, since the jacks can be taken to the vehicle rather than the vehicle being moved over the pits. With a flat depot area and the use of hover pads, vehicles also can be moved anywhere in the depot and then jacked up to working height at the most convenient workstation. A no-pit workshop also reduces the danger of people falling in and being injured; in addition, it is not necessary to install safety railings, which reduce the utility of the pits.

There are also soft requirements for a depot. Because the workforce will need to change clothing at the end of a day's work, showers are an important provision. The depot also may provide a dining area for all staff, including the vehicle drivers. There will be a control center, where operating duties are allocated, services managed in real time, and signaling and electrical power controlled. There will be a storage area for both vehicle spare parts as well as protective equipment and clothing for staff. Finally, it is usual for the system general manager and other senior staff to be located at the main depot, along with training staff and facilities for both maintenance personnel as well vehicle operators.

Ironically, most light rail depots will need a staff parking area, especially because some crews will start early in the morning and others finish late at night and therefore will need to drive to and from their homes. Even if the operator's policy is to employ staff who can use the light rail service to reach the depot or the operator provides transportation to ferry staff, there will always be those who cannot make use of that service and must drive and therefore will need space to park their cars. Visitors to the depot, as well as suppliers of parts, etc., also will need parking space.

4.6 Overhead Line and Electrification

The first tramways were horse drawn, but even though a horse could pull 40 passengers in a tram compared to only 15 by bus, the impact of wars and equine illnesses made operators look for more reliable alternatives to animal power.

With steam railways as the most obvious model in the 19th century, small steam locomotives were tried on various systems (Figure 4.23). The noise and dirt from steam locomotives on public highways were not as acceptable as on remote rail rights-of-way. Other systems of motive power were also tried, including batteries, cable, compressed air, clockwork, and gasoline engines. None was entirely satisfactory, although many cable-hauled systems were built around the world, including in San Francisco in the US, Melbourne and Sydney in Australia, Dunedin in New Zealand, and London, Edinburgh, and Llandudno in the UK. Today, only San Francisco and Llandudno still operate by cable haulage.

It did not take long for operators to realize that electric traction would offer many advantages. The first practical traction system was demonstrated by Siemens in Berlin, Germany in 1882; it used the running rails to provide the electric power. Because this was not a practical option in streets, overhead power line was used by Siemens, first in Berlin. Siemens used a free-running trolley on the top of twin wires ("Troller"). This had serious practical difficulties, especially if more than one line was built, requiring the OHL to split or join at junctions.

The breakthrough came in the US in 1887. Frank Sprague invented an upward sprung pole with a wheel pressed onto and running along the underside of an insulated overhead cable to supply the power to streetcars. The

Figure 4.23 Steam tram operation in Dundee, 1898 (Photo: Dundee City Council, Central Library)

practical demonstration of this was in Richmond, Virginia. The OHL trolley made current collection more reliable, and junctions between lines became practical.

Meanwhile, others were looking for electrical supply systems that were less intrusive than OHLs in the city streetscape. The most widespread in use was the conduit system, which had an opening between the running rails, giving access to an underground channel. A "plow" from the streetcar slid through a continuous slot to contact power conductors. The first US conduit system was opened in Denver in 1885, followed by New York and Washington. A conduit electric tramway was also opened in Blackpool in 1885, the first in the world.

In the conduit system, the electric conductors are placed underground, and the vehicle collects the power through a slot in the road surface, normally between the running rails but sometimes under one of the running rails. This system of power supply found favor in a number of capital cities, including Brussels, Budapest, London, Paris, and Washington, D.C. The last operator to use the conduit system was in Washington, D.C., and it closed in 1962.

In addition to the high cost of excavating and installing the conduit, maintenance costs also were high. Debris that fell into the conduit had to be cleaned out, and if a vehicle came to a halt at a junction, it lost power and was stranded. It had to wait for the next vehicle to push it onto the power rails on the other side of the junction. The conduit system, while practical, was very expensive to build and maintain. Therefore, some cities compromised and used the conduit only in the central area and OHLs in the suburbs. This required changeover points part way along routes, which held up service and were labor intensive. The delays could be severe, especially if there was a shortage of plows used to collect the power from conduit rails for streetcars headed to the downtown.

The second alternative, which used an underground electric cable, transmitted the power to passing streetcars via surface pads or plates. Various methods of activating these, including mechanical and magnetic, were tried, but none proved entirely reliable, sometimes leaving the surface contact pads electrically live. Reports of electrocuted animals and the difficulty of maintaining these systems led to their demise by 1910.

Interestingly, therefore, Alstom has developed a new surface contact system which has been installed in the center of Bordeaux; it opened in 2005. The surface plates are activated by a radio signal from approaching LRVs. At first, this was unreliable, but with more development has been made to work, although it was reported to cost at least four times as much to install as the OHL system. Due to the high cost, this system has been used only in central Bordeaux, with OHLs elsewhere. There is as yet no evidence of a solution to the other problem which beset the earlier surface contact system: surface water, from rain or snow, leading to arcing or short circuits.

Other systems for powering tramways have included internal combustion engines, batteries, and later flywheels to store energy on-board. For various reasons, these have all proved impractical, unreliable, or uneconomic. In the future, with the development of and reduction in the cost of supercapacitors to store power, and with roof-mounted photovoltaics, it is possible that autonomous power systems may be practical and not need an external power source. This, however, will eliminate one major advantage of an external power source: the ability to draw extra current, overload motors, and get more power for climbing hills, etc.

Historically, the OHL became the universal method of electrifying streetcar lines and tramways, but not before the other methods discussed earlier were tried. In the late 19th century, most towns and cities were already overhung with many wires carrying telegraph, telephone, and electric power, and many city authorities were concerned that the additional tramway OHLs would be aesthetically unacceptable. There were also problems with interference from cables and reports that lighter telegraph cables were blown onto streetcar OHLs with dramatic and sometimes fatal results.

The OHL is the most practical and economic system for tramway and light rail electrification. OHLs can be aesthetically pleasing. With the advance of OHL design, they can now be inconspicuous, especially if adjacent buildings are used to support them. Even if traction poles are used, they can be slender and, in partnership with the highway authority, can also be used to provide for streetlights. The Vienna Opera House, a World Heritage Site, is surrounded by the OHL for the local tramway and indeed provides support and thus avoids the need for poles (see Figure 4.3).

There are three major considerations for the OHL system design engineer: electrical, mechanical, and aesthetic. As LRVs became heavier, more power (current) was needed. An external power supply also has one additional advantage. It allows the vehicle designer to use oversaturated motors during acceleration by drawing more power from the OHL than the continuous rating of motors would indicate. Typically, this oversaturation can be about 200% of the continuous power rating. With careful attention to the design of the traction control system, this means that smaller and lighter motors can be used, without overheating.

Compared to the electrical resistivity of the running rails (10×10^{-8} Ω/m) used as the current return path, the OHL (1.7×10^{-8} Ω/m) is at least an order of magnitude less electrically resistant (Figure 4.24). In practice, because of the much larger cross section of the rails, the resistance of the tracks is about 40% lower than that of the OHL. In the UK, the voltage of the running rails normally should not be more than 7 V above earth potential, requiring substations to be about 2 km apart. The voltage drop in the OHL over the same distance is likely to be about 50 V.

To limit the power losses in the OHL, the electrical resistance needs to be reduced. This can be achieved by the choice of conductor, usually copper, copper alloy, or aluminum. There are better metal conductors (e.g., gold), but they are much more expensive and heavier. The second method for

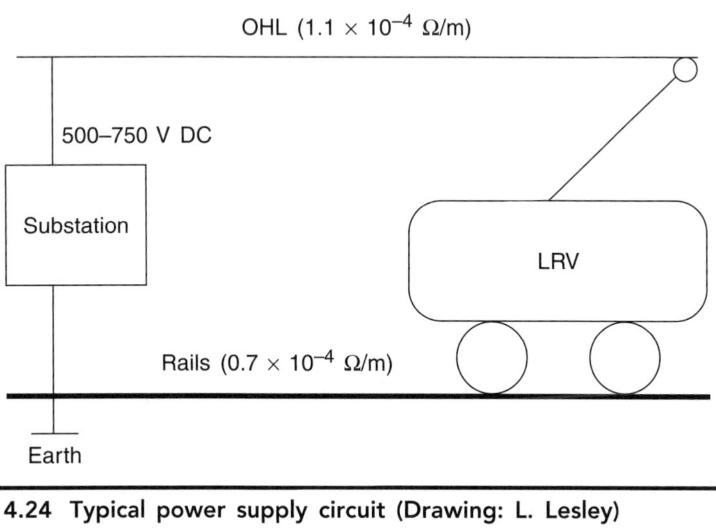

Figure 4.24 Typical power supply circuit (Drawing: L. Lesley)

reducing OHL resistance is to use a larger diameter cable, since the resistance is inversely proportional to the cross-section area, but the diameter is proportional to the square root of the cross section. Therefore, the resistance can be halved with only a 40% increase in the diameter. This minimizes visual impact compared to having a second wire to achieve the same reduction in resistance.

If there is a double-track line, another way to reduce the OHL resistivity is to link the two OHLs to work as one. This does require switching points, where the two can be isolated, when for example work is being undertaken on one track and the OHL of that track must be depowered for staff safety.

A fourth way to reduce resistivity is to have large section underground feeder cables in parallel with the OHL. The advantage here is that the combined resistance is less than the resistance of the feeder cable (Figure 4.24). A fifth way to reduce power losses in the OHL is by having more substations or using low-resistance cable to feed the OHL at shorter intervals (e.g., 800 m).

Resistance of a combined parallel feeder cable for an OHL is

$$\frac{1}{R} = \frac{1}{r_1} + \frac{1}{r_2} \tag{4.2}$$

where R = resistance of the combined system, r_1 = OHL resistance, and r_2 = feeder cable resistance.

Let $r_1 = 50\ \Omega$ and $r_2 = 1\ \Omega$. Therefore:

$$\frac{1}{R} = 0.02 + 1 = 1.02$$

and

$$R = 0.98\ \Omega$$

A possible sixth way for some suburban or segregated lines is the use of railway-style catenary, where the upper cable acts as the feeder to the lower contact wire. Rarely will this be acceptable in city streets. Where there is parallel power feeding to reduce losses on double tracks, appropriate sector isolation makes it possible to maintain operational flexibility for maintenance and repair on one track while still using the other.

Whichever method of feeding power to the OHL is used, the major structural design task is the location of the support system, which will accommodate both the passage of LRVs and weather extremes in terms of temperature (expansion/contraction), wind (sway and horizontal forces), and icing (extra weight and damage to pantographs or other current collector).

Figure 4.25 Span wire wall fixing in Sheffield, 2008 (Photo: L. Lesley)

The OHL is normally suspended from span wires or by bracket arms from poles set into the ground. Span wires can also be directly hung from adjacent buildings (Figure 4.25), which reduces the need for poles and the visual clutter of streets. As an example, the city center section of the Manchester Metrolink was originally going to have 75% of supports directly hung from buildings and 25% from poles. In practice, 75% are from poles, considerably adding to street clutter and giving rise to an ongoing source of criticism.

The pullout force on a building support from double-track span wires is quite low, usually under 1500 N (about 300 lb), which masonry walls can easily accommodate. Historically, a large plate (rosette) was fixed to the wall with four bolts cemented into the masonry. Today, a single "eye" bolt is bonded into the masonry by an epoxy resin or similar high-strength cement. Normally, the building owner is given a small annual payment, and the LRV operator ensures against building damage from the OHL.

Historically, OHL support poles have been sunk about 2 m (6 ft) deep into the ground, usually with a slight slope (about 6 in./150 mm from vertical at the top) away from the track center, to compensate for the pull in weight of the wires. More recently in the UK, heavy rail practice has crept in, using concrete blocks about 1-m cube set into the ground, with the poles bolted on top.

Finding the space for such large blocks is often difficult in the footway or sidewalk, considering the number of understreet utilities. Also, for the stresses involved, the use of concrete blocks is overengineering. A demonstration project for a simplified OHL design was undertaken in 2004 at the Carnforth Railway Centre in the UK. Wooden poles 7 in. (180 mm) in diameter were set 2 m deep into nondescript mixed ground with very satisfactory results. Holes were dug with a small backhoe. The poles were planted and the ground was compacted without the use of concrete. Within an hour of planting, linesmen were able to climb the poles (Figure 4.26).

A variety of OHL poles are used by different light rail systems. The most widely used are made of steel and are tubular and often tapered, but there are also examples of concrete, timber, H-section rolled sections, and lattice steel masts. There are also examples of suitable adjacent trees being used to anchor span wires. In cases where poles are used, they will be about 26 ft (8 m) above the road surface to allow for wire sag, to keep the OHL a safe distance above the road surface, usually about 18 ft (5.5 m), to give enough space for tall trucks (and double-deck buses) to pass safely underneath and across the tracks. In some places, there are reserved or strategic "high-vehicle" routes, where the OHL will need to be 23 ft (6.5 m) above the road and therefore "high-reach" pantographs will be needed on the vehicles.

Today, most light rail systems use pantographs; therefore, the OHL, in addition to following the line of the track, also wiggles from side to side across the track (Figure 4.27). This ensures that there is even wear on the panto-

Figure 4.26 Wooden pole used at Carnforth (Photo: L. Lesley)

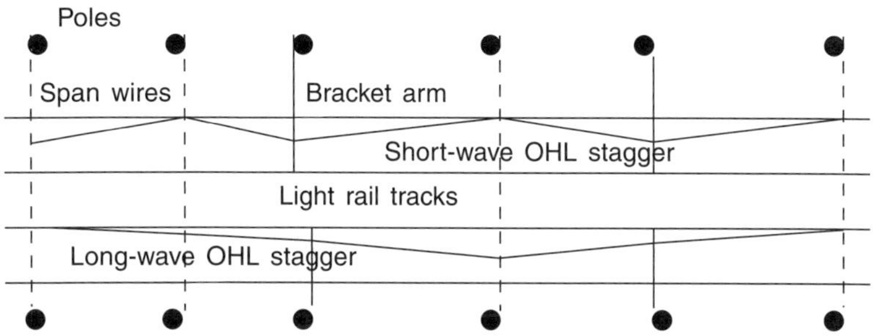

Figure 4.27 OHL stagger for even pantograph wear (plan section) (Drawing: L. Lesley)

graph head. For systems that use a mixture of both trolley poles and pantographs, the pantograph usually will be mounted midway between bogie centers, whereas on exclusively pantograph systems, the pantograph is mounted above a bogie center.

Normally, the OHL is double insulated from its supports. Given the low voltage involved (under 1000 V DC), simple insulators are adequate. Historically, these were ceramic. Today, one of the many plastics with the right mechanical and electrical properties can be used. For the Carnforth demonstration, plastic rope was used as the primary insulator, with wooden poles or plastic collars insulating the bracket arms from metal poles.

Mechanically fixing the OHL directly to span wires or bracket arms creates hard spots where pantographs (and trolleys) bounce, causing arcing and damaging the OHL and current collector. Railways avoid this by using a catenary. This gives an elastic support to the entire OHL contact wire but then needs registration arms to keep the contact wire to the correct horizontal alignment. This is not practical or aesthetically acceptable for street tramways. A new system in use since 2004 at the Carnforth Railway Centre provides a fully elastic OHL suspension and zigzag over track without the use of a catenary or registration arms (Lesley 2006).

Until recently, tramway OHL was not tensioned, as it was set up for average temperatures and therefore slack in hot weather and taut in cold. Tensioning an OHL has many advantages, including temperature-independent OHL height, consistent wear on pantographs, and the ability to run faster without OHL or pantograph damage.

Railway catenary tensioning uses weights or springs to pull the end of OHL cables. This then needs OHL overlaps to maintain a conductor over the whole track. There are few examples of such tensioning in city center streets. Most use untensioned OHL to avoid the clutter and pole strengthening needed to resist the typically 10-kN (1-ton) or greater force. The Carnforth demonstration system, now patented, has developed a completely

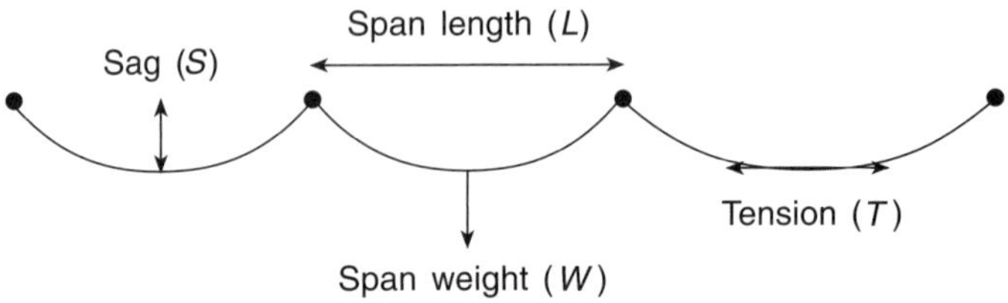

Figure 4.28 OHL tension (vertical section) (Drawing: L. Lesley)

new tensioning system that does not need OHL overlaps or large-end tensioners and places trivial extra forces on tensioning poles (Figure 4.28).

The relationship between the factors involved in hanging an OHL following a catenary curve between supports can be determined from:

$$T = W \times \frac{L^2}{(2 \times S)} \tag{4.3}$$

where T = tension, W = span weight, L = span length, and S = sag. Therefore, if L = 40 m (130 ft), S = 0.2 m (8 in.), and W = 54 kg (120 lb), then:

$$T = 54 \times 40 \times \left(\frac{40}{2 \times 0.2}\right) = 216 \text{ kN } (500 \text{ lb force})$$

If S = 0.1 m (4 in.), then:

$$T = 54 \times 40 \times \left(\frac{40}{2 \times 0.1}\right) = 435 \text{ kN } (950 \text{ lb force})$$

This allows the overhead engineer to consider the option of longer span lengths on straight tracks against higher tensions to maintain an acceptable sag. Thus:

$$L = 50 \text{ m } (164 \text{ ft})$$

$$T = 844 \text{ kN } (1860 \text{ lb force})$$

Curves present another challenge for the design engineer in maintaining the alignment of the OHL over the track: accommodating the dynamic kinetic envelope of the passing LRVs and keeping their pantographs in contact with the OHL, whatever the weather conditions. The tighter the curve radius, the greater the care that has to be taken with the design, since fluctuating temperatures will change the alignment of the OHL.

Figure 4.29 Aligning OHL over curved track in Sheffield, 2008 (Photo: L. Lesley)

This must be accommodated by the careful use of span wires and pull-offs. Figure 4.29 shows an 80-ft- (25-m-) radius curve on city street track. The OHL engineer has used split wire supports to prevent sharp kinks from forming in the OHL, which could bounce the pantograph and cause sparking, thus damaging the OHL and pantographs.

In between the pull-offs, the OHL follows a straight path, so the points of OHL support on curves are usually on the outer edge of the pantograph track. The distance between two supports allows the OHL to migrate to the extreme points of contact on the pantograph, taking into account the pantograph and vehicle sway. Finally, it is worth noting the minimal number of traction poles required by this arrangement and that all the poles are on the outside of the curve, with adjacent poles sharing the middle OHL support.

4.7 Substations and Distribution

The first generation of electric street railways and tramways in the 19th century generated their own power, because electric power supply was still in its infancy and most cities did not have electrical grids. Today, most light rail systems buy their power from central or public generators. This will be delivered to the light rail system at a high voltage, normally 10- to 25-kV three-phase AC. Substations, typically about 2 km (1.25 miles) apart, then transform and rectify the power to the OHL line voltage, usually 500 to 750 V DC.

Substations are normally unmanned and therefore must be remotely controlled, including the ability to switch power on and off and to isolate the substation or sections of track served by the substation, especially in an emergency. There will be considerable protection inside the substation, including high-speed switches in the case of short circuits on the line or inside an LRV. These usually operate in milliseconds. Remotely monitoring the functioning of substations is now standard, using either a dedicated cable network (wire or fiber optics), radio links through the air, or a radio frequency carrier over the OHL.

Stray currents (Figure 4.30) have been an issue in new UK systems, alleged to cause damage to understreet utilities and steel foundation frameworks of buildings. By and large in most countries, stray currents are considered a problem only when actual damage (e.g., steel corrosion) is observed. In the UK, measures to prevent strays have added to the cost and complexity of new light rail system tracks. These measures include the use of Faraday cages under the track, which one senior engineer claimed are ineffective (Snowdon 2007). Recently, track has been relayed by the same engineer on the Croydon Tramlink, omitting stray current protection, on the basis of cost

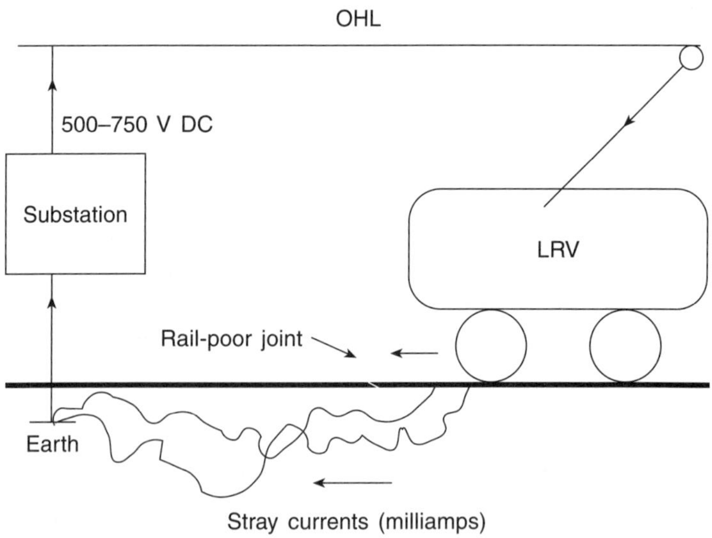

Figure 4.30 Stray current paths (Drawing: L. Lesley)

savings and no observable stray current effects. Two other methods of stray current reduction are available, one systemic and the other location specific.

If the track for the return circuit is not grounded, then there is little scope for currents to stray through the ground. There are ungrounded light rail systems that are as safe as the majority that are grounded. Another way to reduce stray currents is if rails are totally insulated from the ground. A widely used method of insulating tracks, mechanically as well as electrically, is embedment polymers. However, in wet weather, all grounded systems will have stray currents. These result from water lying on the road and over rails, with currents going across the road surface and down drainage channels, etc., following the rainwater.

Location-specific measures include diodes to prevent catalytic corrosion of steelwork. This, however, requires the excavation and treatment of the steelwork. In most of the world outside the UK, this is the form of protection most widely adopted.

Unmanned light rail substations (Figure 4.31) typically with a 500- to 1500-kW capacity can be fitted in small spaces, including basements or other places that normally would not be in active use and even into prefabricated portable containers. Such substations will have an interface with the incoming high-voltage supply, which will include fuses and isolators and may include metering for payment. Transformers and rectifiers are now compact,

Figure 4.31 Portable substation in container (Photo: L. Lesley)

leaving only the DC switch gear to isolate the OHL from the substation or to isolate just one OHL for track or other maintenance and allowing the other track to be used safely for service trams. Safety monitoring equipment also will be linked to the central control, so the supervising engineer can see the condition of all the equipment and take preventive maintenance steps where needed.

From the substation, cables will supply feeder boxes about every 800 m (900 yd) or so, which can also perform OHL isolation functions. These parallel feeder cables will maintain OHL voltage when LRVs are drawing power by reducing the overall resistance of the OHL network. The ducting for these cables can be a considerable expense unless undertaken at the same time that tracks are laid. Cables in these ducts will be specified <3000 amp and will be installed so that there is sufficient cooling to prevent the cables from overheating. This is important because overheating increases the resistance of the cables and therefore the power loss in the system. Overheating is not normally a problem for light rail systems because power drawn by passing LRVs is variable and rarely at maximum current for more than 30 seconds, except perhaps for key feeders in busy city centers.

In some light rail systems, the LRVs regenerate current on braking and feed that back into the OHL. If there is another LRV in the same section drawing current (e.g., during acceleration), the regenerated current can be conveniently used. On a light rail system with an operating service of about 5 minutes or longer, the probability of this is low. In this case, the voltage of the OHL is likely to rise dramatically to more than 1000 V and damage safety equipment and switchgear. Substations will therefore have to be fitted with a way to absorb regenerated current.

Most systems with regenerating LRVs use resistances in the substations to absorb the extra current by creating heat that is then wasted. There have been proposals for fitting energy storage flywheels in substations, which as well as absorbing regeneration currents will also provide voltage stabilization. As of 2011, no such installation has been made. With the development of supercapacitors and their reduction in cost, the same result could be achieved electrically. Nevertheless, the maximum power reduction reported is of the order of 25%, and this saving needs to be offset against the increased capital and maintenance costs of more complicated substations, not to mention the LRVs.

4.8 Construction and Installation

Construction is the most public part of installing a new light rail system and therefore the one that requires the most care in planning and good engineering design in order to minimize the public impacts, including traffic disruption. One of the major criticisms of the construction of the Sheffield, England system in 1993–1994 was that the contractor treated the whole city as a construction site. This caused considerable disturbance to traffic flows and difficulty for frontage businesses to keep open. A similar situation is being experienced in Edinburgh, where track construction began in 2008 and is not due to be completed until 2012.

The largest part of the preparation for the construction of a new light rail system that is not all on segregated rights-of-way will be the roads where the tracks will be installed. This includes, where necessary, the relocation of understreet utility plant that is physically in the way of tracks. In some places where plant will become inaccessible, it may be relocated, although leaving plant in place is often the preferred option of many utility companies, provided arrangements for safe access can be made with the light rail operator.

The use of concrete support slabs, either laid by slip form pavior or cast *in situ* in sections with shuttering, further extends the period of disruption while the concrete cures to the required strength. The only system using precast concrete track panels was developed in Budapest and was widely used in Comecon countries until 1990 (Horvath et al. 1977). It suffered from poor cohesion with the substrate and led to panel rocking and other failures.

Constructing tracks on segregated rights-of-way is easier for the engineer, since there will be exclusive use of the ways and materials can safely be stored on the site without interfering with the public. There are, however, challenges in constructing on continuous and narrow strips of land, including lack of side access. A "moving front" strategy can be adopted, with deliveries over the completed rail sections.

4.8.1 Trackworks

Part of the design engineer's challenge is to construct in the least disruptive installation sequence, including the delivery of materials to the site. Unlike many other construction projects, building tracks in city streets is a very public affair and rarely can 100% occupation be possible or even all the traffic removed, since the city has to continue to function and access to frontage property along the route needs to be maintained.

Many lessons should have been learned from the tramway installation in Sheffield in 1992–1994, where the contractor effectively treated long lengths of the line as a conventional construction site or mainline railway occupation. These lessons unfortunately were not applied to the construction of the Edinburgh tramway in 2008–2012, which has caused considerable public protest, and the opening of the first section will not be until after 2013 at the earliest.

The first lesson is the need for proper consultation with those who require access to frontage property to find out what is needed to maintain their access and how to minimize the environmental disruption. For example, during a critical 2 weeks of the installation in a residential area, the contractor might offer the residents vacations with suitable security of their property while vacant. This both avoids inconveniencing the residents and is a way of thanking them for their cooperation. Not having to work around people can reduce installation costs, which would offset the cost of offering residents vacations. This also should facilitate a faster and smoother installation.

The most disruptive operation is installing the tracks and, where needed, relocating understreet utility plant. Here again, close consultation with the utility companies will pay handsomely, since relatively minor redesigns of track alignments can produce very substantial savings by avoiding or reducing the need to divert plant, and even larger savings can be achieved by avoiding disruption in the highway. Figure 4.32 shows an installation sequence from recent track relaying in Budapest.

Figure 4.32 Budapest embedded-track-laying sequence, 2009: (top left) track panels on concrete strip foundations, (top right) mass concrete poured to rail bottom (note the gauge bars), and (bottom) sand compacted as base for stone block road surface (Photos: L. Lesley)

Inevitably, the tram track horizontal and vertical alignments will not be identical with the existing highway pavement. Traditional urban streets are normally profiled with a crown along the centerline, so that rainwater drains sideways into curbside gulleys. Tram tracks normally are laid horizontally or close to it, so that passengers have a comfortable ride and the wear on wheels, bearings, and rails is equalized between the two sides of vehicles and track.

The first major decision is, therefore, whether to reprofile the highway pavement to the ultimate alignment of the track, before or after installation. Up to now, most track installers have opted for post-track-installation road surface reprofiling (Figure 4.32). The benefit of a preinstallation reprofiling is that it is a simple highway procedure, with many competent contractors willing to give competitive quotations.

Not having to work around rail tracks also makes the resurfacing less complicated. Pavement reprofiling after the tracks are in place is more complex, with an interface between road and rail, and fewer contractors are willing to undertake such work at a competitive price, since the scope for problems is significantly higher.

By reprofiling the road surface to the ultimate track alignment first, the design engineer can undertake quality control in two stages. The first is the highway pavement, where any remedial measures needed are clearly the responsibility of the profiling contractor. The second is with the track installer, who can use the highway pavement alignment as a datum. Thus, any errors in the track profile are the responsibility of the track installer. The installation process also should thus be simplified, since the two operations are separate and in sequence.

Installing tramway or streetcar tracks will depend on the kind of rails used. Grooved girder rails, such as Ri60 (Figure 4.33), are usually about 7 in. (180 mm) high and bottom supported. They require a foundation underneath the rails to provide support. Recent new systems in France have effectively used mainline railway track designs with a crushed rock ballast base and the rails clipped in place on cross-ties. This means that a width of about 24 in. (600 mm) of the road has to be excavated to a depth of at least 40 in. (1000 mm).

For extensions to existing systems, often the original rail profile is not available, and therefore a transition between different profiles is needed. Where LR55 rails (Figure 4.34) extend existing "Phoenix" grooved rails like Ri60, then a radical transition is needed (Figure 4.35), not only to match the different profiles but also to maintain rail stiffness across the transition.

Many North American and other European systems use a concrete slab between 300 and 500 mm thick with the rails clipped or bonded to it; this requires an excavation depth of between 500 and 700 mm. The newer LR55 system, developed by the author, uses the strength and stability of the existing road pavement base with shallow foundation prefabricated concrete beams/troughs only 7 in. (180 mm) deep and 13 in. (340 mm) wide, set into

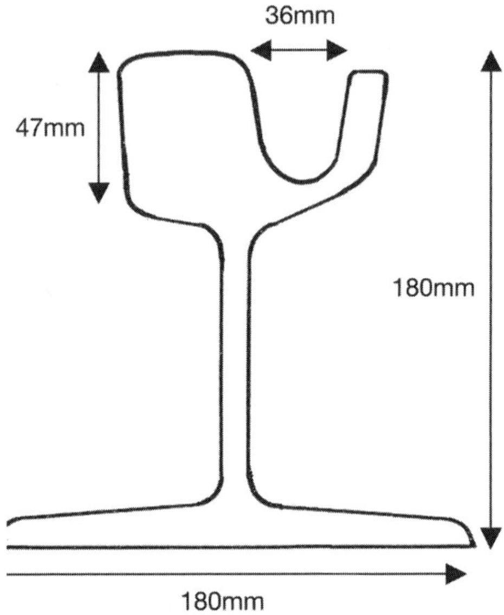

Figure 4.33 Ri60 rail profile (Drawing: L. Lesley)

narrow trenches 8 in. (200 mm) deep and 14 in. (350 mm) wide (see Figure 4.10). This minimal excavation is possible by the use of a top-supported rail only 80 mm deep (see Figure 4.34).

Probably the most complex part of this process is the installation of turnouts (switches and crossings), where tracks divide, merge, or cross over each other. The design engineer is responsible for ensuring that the track fabricator has, as early as possible, the horizontal and vertical track plans. This will enable the fabricator to preassemble the track components off-site, to

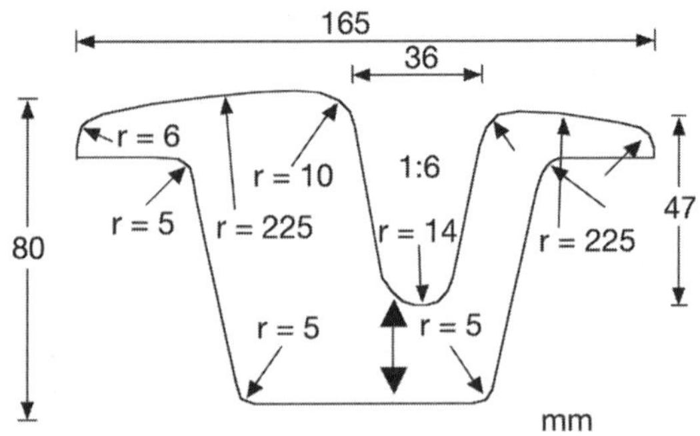

Figure 4.34 LR55 rail profile (Drawing: L. Lesley)

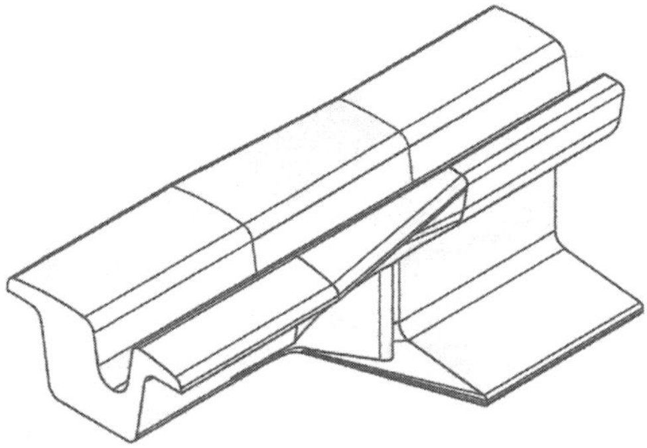

Figure 4.35 LR55/Ri60 transition rail detail (Drawing: David Turner)

ensure that the fit to plan is accurate and to make any adjustments prior to assembly on-site.

Traditionally, LRVs have traversed switches and crossings at low line speeds, which means that passengers are subjected to further braking and acceleration, and power consumption is increased. The reason for this is that street track switches usually are short radius, about 66 ft (20 m), without transitions (Figure 4.36); thus, for the comfort of passengers, limiting hori-

Figure 4.36 Embedded "scissors" crossover in Budapest, 2009 (Photo: L. Lesley)

zontal and centrifugal jerk rates means slowing down, often to as slow as 3 mph (5 km/hour).

New ideas for switches can avoid both the potential discomfort of short-radius curves as well as the need for trams to slow down unnecessarily. This can be achieved by the use of "presorting" points. Here, the switches are set back some distance from the junction, 160 ft (50 m), using large-radius turnouts >650 ft (200 m). Two tracks are then interlaced and run side by side until the junction. At the junction, the diverging track can include a proper horizontal transition with only a simple crossing, where derailments are practically zero. There is a second benefit to presorting junctions, where LRVs have traffic signal priority, in that it makes coordination with the traffic signals easier, thus providing a "green wave" for trams, the norm in many cities.

Where there is not enough room for a presorting arrangement, use of spiral curves across the switch blades with radii greater than 100 m (325 ft) and tightening to 25 m (80 ft) will be more comfortable for passengers and less damaging to track and vehicles. Here, close cooperation between the design engineer and trackwork fabricators will pay dividends in the form of a lower maintenance requirement. Where there is not even room for this, then an initial 25-m radius followed by a straight connection to the second track, with a reverse 25-m radius, will achieve a similar comfortable result.

Historically, embedded road switches and crossings have created additional and often annoying wheel/rail noise, particularly as LRVs are flange guided around short-radius switch blades and then pass noisily over the crossing gap. The use of presorting points means that one source of noise is reduced, as well as reducing switch blade wear and the need for maintenance. The second noise source common in railways, the crossing gap, was solved more than 100 years ago by the "silent" crossing (Figure 4.37), where the bottom of the groove is raised at the intersection of the rails so that wheels are momentarily flange, rather than tire, supported.

A variation of this is where the groove on the parallel rail is also raised, so that both wheels are flange supported, thus reducing any extra torsional forces in axles when wheels are solidly attached to the same axle, although this source of stress is small compared to traversing short-radius curves. Historically, there is little evidence of any problems arising from this.

Most continental European tramways have had "silent" points or turnouts for more than 100 years that give satisfactory service and are environmentally benign, in addition to being good neighbors in terms of the lower noise. The first generation of UK tramways all had silent points, unlike the second generation, which seem to use the railway practice of noisy open crossings. This is a specific example where the design engineer faced with a technical problem would do well to first search the literature (or Internet) to find out

Figure 4.37 Crossing raised grooves (Photo: L. Lesley)

if it has previously been faced and solved, as such research might save significant time, cost, and heartache in the design and installation of trackwork.

While the rails are being laid to make up the tracks, the return current cables will be connected to feeders and substations, to ensure that rail joints not welded are electrically continuous. Similarly, vehicle detection equipment will be installed, as well as the cables for the control of switch motors.

Completion of the highway pavement is the final part of embedded track installation. The easiest way is the use of tarmac or similar flexible pavement materials that can be rolled between and outside the rails to be level with the top of the rails. This gives a smooth road surface and a good waterproof

Figure 4.38 Example of segregated open light rail tracks in Blackpool (Photo: L. Lesley)

seal against the rails. Some cities use stone blocks for the road surface, which are sealed with tar, bitumen, or other mastic. The advantage is that they can easily be removed for track maintenance, but they provide an uncomfortable ride for other road vehicles, although stone blocks might make sense where tracks are exclusive to LRVs to deter trespass from road vehicles.

For segregated tracks that are not embedded in highway pavements, normally the choice of track form and construction is straightforward. This is because conventional flat-bottom rails simply supported on sleepers/ties are the most widely available and therefore lowest cost option (Figure 4.38).

After clearing the right-of-way, and protecting any cross routes to property and other requirements, the first stage will be the removal of any topsoil. If the tracks are subsequently grassed, some of the topsoil can be kept for this purpose. The rest must be disposed of. In urban areas, normally there is a market for topsoil for gardening and horticulture, which might generate a little income to offset part of the cost of installing the tracks.

On reaching the subsoil, drainage must be provided, since waterlogging is one of the common causes of tracks failing. This can be done by grading the formation into side drainage ditches or incorporating perforated drainage ducts along the line of the tracks. A geotextile barrier is placed over this to keep the subsoil and ballast separate. Then a ballast of crushed aggregate about 1 in. (25 mm) in size is laid to act as an elastic base for the tracks.

Normally, prefabricated panels of track 60 ft (18 m) long are brought to the site, laid, and aligned. The rails can then be welded *in situ* into long strings or prewelded long rail strings can be brought to the site, to replace the existing rails. Once the rails are at the correct line and level, additional ballast is laid between the sleepers to anchor the tracks laterally (Figure 4.38). If a softer appearance is desired, another geotextile sheet is laid over the ballast and the topsoil is spread and seeded (see Figure 4.15). The geotextile prevents the topsoil from contaminating the ballast, because its elastic properties depend on the ballast remaining clean.

4.8.2 Overhead Lines

In the last 125 years of electric tram and light rail operation, there have been numerous attempts to supply power to vehicles without the need for OHLs. Of these, the conduit system, with a continuous slot between the running rails and electric supply rails under the street in a hollow duct, was the most widely used and most reliable. To avoid the need for wires above city streets, conduit tracks of great complexity, cost, and operational difficulty were installed in many cities, including Brussels, Budapest, London, Paris, Vienna, and Washington, D.C. The final operational system, in Washington, D.C., closed in 1962, 10 years after the last conduit-powered tram ran in Europe (London). Other systems, including attempts to place conductors on the road surface, all failed for technical, safety, or cost reasons. Recently, a new surface contact system was demonstrated in Bordeaux by Alstom, and Bombardier has developed another system that relies on inductive transfer of power between understreet cables and inductors on vehicles.

The OHL, however, remains the most reliable and highly developed power supply system, and the recent application of industrial design approaches means that OHL can be aesthetically acceptable even in World Heritage cities like Vienna. OHL is economic to install and maintain and is not proprietary.

Installing OHL in city streets requires almost as much planning as the trackwork. The function of the OHL is to safely supply electric power to LRVs, and at the same time it should be aesthetically pleasing and environmentally unobtrusive. The preplanning survey will include locating and identifying street furniture such as lampposts, mailboxes, phone booths, etc. that should be avoided or may need to be moved and suitable buildings that can be used for the direct suspension of the OHL (see Figure 4.25), thus avoiding the need for traction poles. Faster and heavier LRVs draw large currents, so they need bigger supply cables if power losses are to be minimized.

Designers should use an elastically suspended OHL. This avoids hard spots where pantographs or trolleys bounce, causing electric arcing that will damage both the OHL and pantograph. The ultimate elastic suspension is a full catenary. For off-highway locations, this might be the most appropriate solution, especially because the second cable acts as a feeder for the higher current drawn when vehicles accelerate.

In city streets, however, catenary is normally unacceptable due to the visual impact and increased clutter of the streetscape, although cities like Manchester and Sheffield use twin cables suspended together side by side as a compromise to supply the high currents needed, with an acceptable OHL resistivity.

The science of engineering design for the installation of street traction poles has nearly been lost in the US and UK during the 50 years since streetcars and tramways were abandoned. The horizontal forces exerted by span wires that support the OHL are of low magnitude, typically about 1500 N (about 300 lb force). This can be compensated for by erecting poles with a slight outward lean (about 6 in./150 mm at the top). The weight from the span wire will pull the pole inward and make it vertical.

The same logic applies when bracket arms (Figure 4.39) are used instead of span wires. The downward weight imposed on the bracket arm by the

Figure 4.39 Cable-stayed side bracket arms in Carnforth (Photo: L. Lesley)

OHL will produce a turning moment balanced by the bending of the pole back to vertical unless center poles are used (Figure 4.38). In this case, the two OHLs balance each other and the net turning moment on the poles is zero.

Planting OHL traction poles was historically achieved by auguring a hole about 6 ft (2 m) deep, slightly larger than the pole base diameter, and then using a cement/concrete slurry to seal the pole in place in the ground. After a short period, poles are firm enough to string the OHL. The newest UK tramways in many cases use traction poles significantly heavier and stronger than needed and bolted to the top of foundation blocks, which is more appropriate for main line catenary. Bill Gibson, the recently retired engineering director of Blackpool Tramways, with continuous electric operation since 1885, tells what happened when the OHL designer for one new UK system visited Blackpool to inspect the OHL. He looked up and is reported to have said, "according to my calculations that is impossible." The Blackpool engineer replied that this design had not failed in more than 100 years.

Recently (2004), experiments were undertaken at the Carnforth Railway center to determine how strong simply planted traction poles can be. The poles were erected by United Utilities, the local electrical power supplier. Plain wooden poles 30 ft (9 m) long and 7 in. (180 mm) in diameter were used. Auguring turned out to be impractical because of the extremely variable nature of the ground, with quarry rubble at one extreme and waste steam locomotive fire ash at the other.

The holes were excavated using a small backhoe (mini JCB). Once a pole was in place, the hole was backfilled and compacted with the backhoe. Within an hour, most poles were stable enough for an installer with spiked boots and belts to climb up them and attach the bracket arms (see Figure 4.23). Where the ground was especially weak, a cementatous ground reinforcement was added to the backfilling and overnight created a stable pole.

Historically, wooden poles and indeed trees have been used to support OHL. Most poles, however, are tubular steel. The kinds of poles that are used will depend on economic, engineering, and political considerations. For street traction poles, careful planning is needed to locate the optimal places for poles to support OHL to the correct alignment for track geometry, vehicle swept path, and pantograph position.

Existing street furniture, including lighting columns, also will have to be considered. Where the lighting columns are substantial, they might be used for the OHL, but they could bend inward, in which case they would need to be replanted. Alternately, the new traction poles could be used to suspend lighting fittings to reduce the clutter of poles along the street (Figure 4.40). In any case, careful consultation with the highway authorities will be needed to determine placement of traction poles and also to ensure that road signs

Figure 4.40 Shared traction and lighting poles in Liverpool, 1938 (Photo: Liverpool City Engineer)

are not obscured. In conservation and areas of special historic or architectural interest, the planning authority also will require close consultation, so that poles are sympathetic to the local area.

To protect the general public in the vicinity during installation, it will be necessary to temporarily fence off parts of the street while pole holes are being dug. Depending on the location, a series of holes can be prepared, and a truck loaded with poles can proceed along the street and plant them in one pass, using a Hiab or similar attached crane or hoist. This would minimize local disruption as well as speed up installation and control costs.

In many city streets, stringing OHL from adjacent buildings is a more environmentally pleasing solution (see Figure 4.25). Older masonry, brick, or stonework buildings are almost always strong enough to support span wires. Historically, elaborate plates (rosettes) were attached to building walls, from which the span wires were fixed.

Today, the use of synthetic resins means that structural bolts are unobtrusive and economical, although without architectural character. Many modern buildings have steel or concrete frames, with lightweight claddings. The claddings are usually unsuitable for attaching span wires. However, where the building frame is at the surface or can be accessed, it will provide a suitable support for a span wire.

If fixing OHL span wires to buildings, agreement of the building owners is required. This is another area where early discussion is important to generate good public relations. Getting building owners to agree to OHL fixings is preferable to using compulsory powers. Compulsory powers will generate negative publicity and even outright public resistance, which is what happened to the proposed tramway in West London. The political party that was promoting the system was voted out of power in two of three districts involved and the project was abandoned.

Installing building fixings as well as bracket arms on traction poles normally is done using a "cherry picker" or similar mobile high-reach platform (Figure 4.41). One was used in the Carnforth project, described above, to install the bracket arm assemblies as well as the OHL. Self-powered work platforms allow installation crews to proceed quickly and safely along a street, installing the OHL fixings on a moving front basis.

The actual location of the OHL will depend on the track geometry and LRV characteristics. For new systems, the OHL will follow the centerline of the track with a horizontal "zigzag," so that the pantograph contact strip wears evenly. Traditionally, the zigzag has been "short wave" between adjacent span wire attachments (see Figure 4.27). Experiments in Carnforth with "long-wave" zigzags, between every fourth support (about 660 ft/200 m),

Figure 4.41 "Cherry picker" used for OHL installation (Photo: L. Lesley)

achieved the same result but placed less sideways tension on traction poles and span wires. On double tracks, the OHL zigzags in opposite directions, so that the lateral forces on the poles or span wires cancel each other.

On curves, the OHL will need pullouts to maintain the centerline within the geometric tolerance of tracks and vehicle swept paths. These pullouts can be supported on span wires parallel to the track to minimize the number of traction poles used or from supplementary building supports (see Figure 4.29).

Tensioning tramway OHL can be a "black art." Some systems do not have active tensioning and operate by self-tension created by the weight of the OHL, with variable sag depending on the ambient temperature (Equation 4.2). For systems that use active tensioning, this is relatively simple for off-street tracks. Tensioning weights, springs, or hydraulic rams are well established and readily available. The calculations needed are easily applied. Tensioning on-street is more difficult, since existing tensioning systems are bulky and not conducive to a simple aesthetically pleasing appearance. This is one reason why tensioning is rarely used for streetcar systems.

A new simple and unobtrusive tensioning system has been developed which can be incorporated inside hollow traction poles and is easy to install. This patented system, available through a license to all competent OHL installers, has been tested since 2004 on the Carnforth test installation.

Installing the OHL is the trickiest part of the process, since ideally it should be done right the first time, with a minimum of subsequent adjustments. The OHL is installed using a moving front process. It starts with anchoring the working end to either preinstalled OHL or a traction pole or similar fixing point. In attaching subsequent fixings to span wires or bracket arms, the correct horizontal location above the track, height above the road, and working tension must be achieved simultaneously. This is the part of the process that requires the most skill. If a new team is doing the installation, an off-street practice location is recommended to gain experience before working on the actual installation.

Normally, a template placed on the tracks will show the centerline and compensate for any cant or change in horizontal track alignment. It also will mimic the LRV and its pantograph. Whereas a supervising engineer can be present for the other aspects of OHL installation, the space available in a cherry-picker platform does not allow for this in the final clipping of the OHL. As the OHL will be supplied in about 1000-yd (1000-m) lengths on a drum, maintaining working tension and alignment will be critical to the final alignment and performance of the OHL.

In the Carnforth project, the OHL was first run out over snatch blocks on alternate poles over the length of the track (Figure 4.42). This was less problematic to the final clipping than having to run out the OHL and, at the same time, clip in position to the right tension. In a street tramway, with

Figure 4.42 Running out the OHL (Photo: L. Lesley)

curves and corners, further complications will have to be accommodated to ensure that the correct alignment is maintained. This might be done as it was at Carnforth with temporary snatch blocks or temporarily looped over span wires or bracket arms. Either way, this is a two-pass process in which the final alignment, height, and tension are achieved in the second pass. As the clipping progressed at Carnforth, a mechanical tensioner was used to provide a tension of 3 tons (30 kN) in the OHL.

The final check is to tow an LRV with its pantograph raised to confirm that the pantograph is in continuous contact with the OHL and that the contact pressure is uniform along the entire line. Minor adjustments to correct small out-of-design variations can easily be made from a mobile work platform which moves with the LRV.

4.8.3 Stations and Stops

In engineering terms, the track and OHL are the major design and installation challenges. For passengers, however, the most important aspects of the system are the station or stop where the journey starts or finishes and the vehicles on which they ride. These are the main and direct passenger interfaces when using a light rail system. It thus makes good sense to have imaginative station designs that are constructed to the highest standards.

Standard design also is important for local residents who may not be passengers. Stations are a permanent part of the townscape and will be a neighbor to dwellings or businesses. They affect the efficiency of the light rail system and add to the civic status of a town. Stations built to high design standards also act as advertisements and ambassadors for the light rail system and will provide corporate identity.

A station with a good image and ambiance is especially important for passengers who would normally travel by private car. A journey by light rail can create anxiety due to the unfamiliar. Drivers who are used to getting in their cars will be especially anxious when they have to wait at a station. The great attraction of private car travel is the perception that it is always available and a journey can be completed nonstop. People endure or overlook traffic jams because of the benefits and convenience of personal travel by car.

Stops where passengers interchange can be particularly unnerving, even with excellent signage, since people do not want to appear to be novices or, worse, get on a vehicle going in the wrong direction. While most of the above is particularly relevant in the design stage, construction and finishing must be carefully supervised to ensure that, for example, signs are put in the right place and face the right way.

With the need to provide access for passengers with disabilities now almost universal, level boarding and alighting from the vehicles is necessary. This will facilitate faster and safer boarding and alighting for passengers in wheelchairs, parents with children in carriages, as well as the elderly who find steps difficult. The 95th percentile of adults can manage an 8-in. (200-mm) step height, but 100% can manage 2 in. (50 mm), which is the usual operating margin of tolerance between platform and vehicle floor.

There are two schools of thought on how to achieve this. One uses high platforms (900 to 1000 mm/36 to 40 in. to the top of the rail) and high-floor vehicles (see Figures 4.2 and 4.43). Increasingly, the norm is to use low-height platforms (typically 250 to 350 mm/10 to 14 in. to the top of the rail) with low-floor vehicles. The option adopted will depend mostly on local geography. This is especially important if a significant part of the system will be in narrow city center streets, where fitting high platforms would be a major planning challenge and might disrupt the normal economic and social life of the city. A final consideration is, of course, that new low-height platforms cost less than high platforms. More low-platform stations can be built within a given budget, further improving access to the light rail system. Shorter walking distances mean door-to-door journeys will be faster and therefore the system more attractive.

There are a range of options in constructing stations, from building everything *in situ* to a large amount of prefabrication with on-site erection. The latter makes sense when station design is largely standardized. On the other hand, if every station is to be different, then on-site construction may be the most economic solution. Whichever approach is adopted, building stations in

Figure 4.43 High-platform light rail system in Manchester city center (Photo: Greater Manchester Passenger Transport Executive)

busy city streets will create a disturbance. The design engineer can reduce this by minimizing the construction time and by careful traffic management. If the end result is a beautiful addition to the street scene, then, like any other attractive building, the station will soon be accepted as an important part of the civic infrastructure. A beautiful and permanent station will also be a reminder of an efficient transit system. Residents will forgive the temporary disruption during construction in exchange for the long-term amenity provided. A visually attractive station will help to build a regular and loyal patronage, provided maintenance and regular cleaning keep it that way.

In addition to physically installing the station, the supervising engineer first needs to check that the alignment of the tracks is correct. Alternately, a station can be built first and the track engineers can fit the tracks to the station. The relative alignment of platforms and tracks is crucial to the safe and efficient operation of a light rail system. This may mean some minor redesign to fit in with the street and carriageway.

Stations also need utilities, such as electricity to power lights and signs. In some locations, station roof space might be used for photovoltaic solar panels to provide the power needed for these functions. In suburban locations, wind generators may be installed at stops to provide this power. It also will be necessary to provide drainage, as well as access to manhole covers located under platforms. An important part of system design is now passenger information, about both the entire system and departure times from a particular station or platform. Good and clear signage is important at every

Figure 4.44 Iconic light rail station in Strasbourg (Photo: L. Lesley)

station on the system, since passengers boarding as well as those alighting will need to be sure that they are at the right station and will need information on how to reach their final destination.

Creating a safe working site will be the first outward sign of a new station being constructed. Depending on traffic management arrangements, ideally the whole road, but at least half of it, will be closed off to allow the foundation to be laid and prefabricated units, where used, to be assembled on-site. If the station is a special design (Figure 4.44), then construction will start from the ground up, which means the road will be closed longer.

Arrangements for access to frontage property will be important, as will be safe delivery of building materials and storage at the site. Normally, construction will only be done during the normal working day (e.g., 7:00 a.m. to 7:00 p.m.). This is particularly critical if a station is in a residential area, as people do not like being disturbed at night. Site supervision and liaison with local people will be important roles for the light rail design engineer.

Part of station installation is the protection of completed works until the system can be opened, when passengers and the tramway's own security system will be able to provide proper surveillance. This may mean delaying the fitting of key elements (for example, passenger information screens) until shortly before opening, although all ductwork and wiring will already be installed and should have been tested to ensure circuit continuity and commissioned to make sure it is working as intended.

4.8.4 Traffic Management and Priorities

A new light rail system should attract commuting and other journeys presently being made in motor cars. It is logical, therefore, that LRVs should enjoy a measure of protection from traffic congestion, including priority at road junctions. The planning and engineering of that will require careful consultation and computer simulations to make sure that congestion is eased and that any spare capacity created is not filled by traffic that is presently suppressed by congestion.

Before that, during construction, temporary traffic management measures will be needed to provide a safe environment for the construction teams and for local residents. Clearly, temporary traffic management will be implemented on a moving front basis as track installation progresses. Where possible, this temporary traffic management should become the final traffic management. The main reason for this is that people are inherently conservative and do not like unnecessary changes.

Experience shows that no matter how extensive the publicity and advance warnings before construction begins, drastic traffic management measures, which significantly reduce automobile route options and road capacity, are likely to create a negative public reaction. This almost certainly will have local political impacts. Therefore, during the construction phase, the engineering team should aim for the ultimate traffic management system, since the fewer road network changes there are, the better. Maintaining access to frontage premises during and after construction is a legal requirement. Discussing these requirements beforehand with the parties that will be affected is sensible consultation and a practical form of good manners.

The ultimate traffic management arrangements should be designed to ensure that LRVs are never stuck in traffic congestion. There are two main ways to achieve this. Where space permits, the light rail tracks can be physically separated from road traffic (see Figure 4.14). The use of existing medians in dual carriageways or wide verges can be an economic and operationally beneficial solution (see Figures 4.15, 4.38, and 4.40).

Separation also can be achieved by displacing traffic from a lane and making it exclusive to trams (and in some places also buses). In the Netherlands, such "trambaans" are raised about 100 mm (4 in.) above the rest of the carriageway to reinforce this physical separation. Exclusive and paved tram lanes can, of course, be used by emergency services and therefore help to reduce response time to incidents. It also should help to ensure that emergency services will support this kind of priority for light rail.

The second way to achieve separation is by the use of time allocation. Typically, light rail operates at a frequency of about every 5 minutes, with a junction passing time of about 20 seconds. This means that for the other 4 minutes and 40 seconds, road space can be used by other traffic, provided the road is clear when the next tram arrives. This "green wave" can easily be achieved by coordinated traffic signals linked into an area traffic control network. The network monitors road traffic flows and a green signal is triggered by an approaching LRV. At junctions, a preempted green signal for a tram would cut short a side road phase, which can be compensated for in subsequent signal cycles by reallocating the tram phases not needed. This keeps the main roads and tram routes flowing and confines congestion to the side roads.

Probably the most advanced traffic priority system is in Zurich, where all traffic lights automatically give a green signal to an oncoming tram (or bus). Contrary to popular belief, this actually speeds up all traffic, since trams and buses accelerating from a stop at a traffic signal take longer to clear the junction than crossing without stopping, therefore releasing more capacity to other traffic. While the Zurich system speeds up all traffic, it speeds up public transport significantly more and therefore has resulted in a modal switch from car to tram (and bus). This is a win-win situation.

How can this level of junction priority be achieved? One approach is to program a tram green phase on every signal cycle that coincides with the timetabled crossing of a tram. If tram services ran 100% reliably, this would be the simplest and cheapest option. Experience shows that 100% operational reliability is impossible. Zurich and other European cities use tram (and bus) detectors so that an approaching tram will immediately call up a tram phase in the signal cycle, and once the tram passes, there will be extra time to compensate the other (nontram) phases.

These detectors can be simple magnetic loops in the road, tuned for long vehicles, and thus will not be triggered by cars. This is the system that has worked for more than 40 years on the Runcorn Busway; it gives buses running at 40 mph (65 km/hour) uninterrupted crossing of road junctions. More often, a beacon on the tram transmits an identification signal, both to trigger a green phase and where there is a tramway junction to change the points, calling up a turn filter phase so that the maneuver can be undertaken without stopping. The ability to identify individual vehicles also can be used in a regulation and management system, where trams running early can be made to wait for a green, whereas trams running on time or late will always have an immediate green.

The mechanics of traffic management during construction will be different from those required after the opening of the new light rail line. During construction, temporary and moveable barriers and signs will be used. These will prevent vehicles or pedestrians from tripping or falling into excavations. They also will direct traffic to alternate routes, enabling the construction sites to largely be free from traffic. Figure 2.8 shows a pedestrian crossing at the track slabs, with orange fencing, to enable access to the footway adjacent to shops.

The permanent traffic management arrangements will require signs, normally put up during the construction phase but covered up. This means that when the new system is ready to open, the signage can be unveiled as appropriate. Traffic signal coordination will need more careful preparation and also may have to be fine-tuned when opened. Experience in major citywide traffic management shows that drivers take about 2 weeks to adjust to new traffic signal settings (Matsoukis 1980). Historically, network traffic signal plans were calculated off-line, working on a fixed cycle time, although with phase splits changing throughout the day to reflect varying traffic flows.

Control computers are now nearly able to optimize a traffic-signaled network in real time. In practice, because traffic flows are predictable on a daily and weekly basis, suboptimization with local detection and control is often used. This makes giving priority to LRVs at road intersections easy. While there is considerable international standardization of road traffic signals (red, yellow, and green lights arranged vertically with red at the top), there are no similar international standards for light rail signals in streets. These will therefore reflect local operating practice.

4.9 Commissioning

Before a new light rail system or line can be opened for passenger traffic, there will need to be a period of commissioning, training, and approval for operation. There also may be a period when no fares are charged, to allow passengers to gain experience in using the system. A free trial is common practice for commercial products and, if nothing else, is good public relations for the light rail promoter. The first stage of commissioning is the static testing of all equipment. Commissioning almost certainly will be undertaken in parallel, rather than sequentially, unless there are secondary relationships.

4.9.1 Electrical Supply and Overhead Lines

Commissioning the power supply to the OHL and the operation of substation switching will ensure that a safe supply of power is available. Tests will include leakage currents, response times to switch-on and -off commands, and the ability to switch out substations while continuing to supply power from adjacent substations. At this stage, emergency procedures can be practiced in response to short circuits, vehicle crashes, and other events where the OHL might present a danger. Switching off power automatically in such an event should be accomplished in a few milliseconds.

Track continuity between one end of the system and the other is vital to identify any broken rail joints or poor return current connections to substations that will create stray currents in a grounded light rail system. As the stray currents involved are normally in the milliamp range, ultrasensitive test equipment will be needed. Repairing any broken rails before service starts will have a high priority, because even if the other rails can provide a circuit to the substation, the risk of another rail break would create havoc.

Figure 4.45 Sand blown across tracks in Blackpool (Photo: L. Lesley)

Similarly, insulation of the traction poles and other street furniture such as lampposts, mailboxes, traffic signs, etc. is important. More than 100 V normally is fatal, so the OHL voltage of 750 V must be carefully tested, especially in very wet weather, when running rain could provide alternate current paths, including making traction poles or other metal street furniture live.

Normally, LRVs should be delivered fully tested. For completeness and as part of the static testing, vehicles can be tested in the depot. The first test is to ensure that, under all operating conditions, there is never a short circuit within the vehicles. If such a fault occurs, then the local substation should immediately switch off the power to the OHL. A second test is to ensure that the vehicle body does not become live if the current returns fail, for example as a result of the tracks being blocked by insulating material such as sand (Figure 4.45), or the wheel return bushes fail.

4.9.2 Trackage

Tracks in streets affect two classes of users: LRVs and other road vehicles going along or across tracks. For other road vehicles, clear lane markings and signs are the best way to ensure compatible interfaces, assuming that contractors have laid tracks compliant with the road surface, which is especially important for pedestrians crossing the tracks.

The first stage of track commissioning is gauging to ensure that tracks are compliant to the gauge tolerance. The second stage is the geometry both along the tracks (alignment) and across the tracks (level), both of which will affect the safe speed of operation and the comfort of passengers, especially standing commuters. The safe ergonomic forces that passengers can accommodate are well established. Comfortable accelerations are also known. The curvature of tracks and the planned operating speed of vehicles will determine the safe and comfortable lateral accelerations. This is determined by the equation for the lateral acceleration force (F) of a body (mass M) at a speed (V) going around a curve (radius R):

$$F = \frac{MV^2}{R} \tag{4.4}$$

Where the maximum centrifugal comfortable acceleration = 1 m/s^2 (~3 ft/s^2), the relationship between curve radius and maximum speed can be plotted (Figure 4.46).

Commissioning will include public education, especially where there has been no recent rail operation on roads. This is particularly important for cyclists, to avoid accidents due to bike wheels getting stuck in the track groove, the cause of an early fatality on the then new Sheffield Supertramway in 1996.

For LRVs, the correct operation of points and signals is important for efficient and safe operation. This will be demonstrated by a vehicle operated

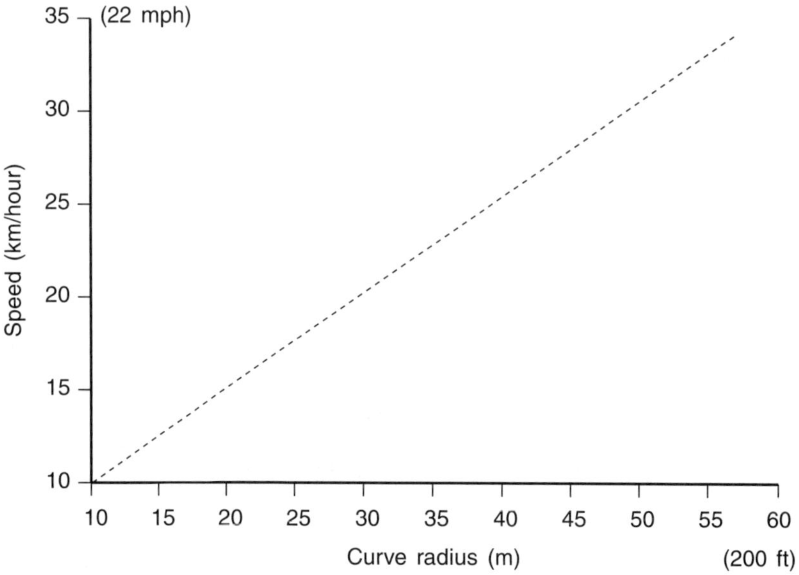

Figure 4.46 Comfortable curving speed (Drawing: L. Lesley)

at low speed and the driver trained to ensure that the right route codes have been used. Should a malfunction occur (e.g., wrong point setting), the driver has time to stop before damage is done to the vehicle or trackside furniture.

4.9.3 Stations

The team that designed and builds the stations should be the first to commission, which means that all systems are working as specified and the structure has been built fully as designed. At this stage, any misfunctioning works need to be identified. In practice, the first place that passengers experience a light rail system is at a station or stop while waiting for a departure. Safe and easy access to the station, including crossing any road carriageway to reach the platform, should be the designer's aim. To check that this is the case, "mystery shoppers" can be used; these are volunteers who undertake consumer testing. They should be neutral, professional, and help to identify any problems not noticed before. Such problems can include the legibility of signs at different times of the day and the evenness and grip of the platform surface under different weather conditions.

Similarly, groups of people (e.g., retirees) can be recruited to board and alight from an LRV at the station. This will allow the flow of people on the station platform to be monitored. It also will allow emergency evacuation procedures to be tested for both LRVs and the whole station area. Pinch points where there is a tight space or difficult area can be identified and measures to ease them retrofitted.

Most importantly, commissioning also should also include orientation and direction signs for passengers to navigate the system. It may be prudent to build one of the stations early in order to obtain consumer feedback, rather than having to retrofit modifications later to all stations. All remedial work identified in this process must be completed before the formal opening of service.

4.9.4 Vehicles

After the stations, the second most important part of a light rail journey for passengers is the ride in the LRV. Despite a lot of preplanning, mock-ups, and ergonomic studies, trying out the light rail system with real passengers will point up areas where small adjustments might make a big difference to operations. Similarly, changes in operational management might make a big difference in the way passengers use and enjoy, rather than endure, the service. Part of the commissioning will include emergency evacuation following simulated accidents. This also will involve emergency services (police, fire, and ambulance).

While a lot of minor accidents (e.g., passengers tripping) occur at light rail stops, major accidents usually occur away from stations, often in junction areas. This has certainly been the experience in Manchester, Sheffield, and Croydon, where road vehicles that disobeyed traffic signals collided with LRVs during their priority use of a junction. In the most serious case, a heavy truck crashed into the side of a Manchester LRV, derailing and pushing it sideways. A police car also was badly damaged in San Diego when turning left across the light rail tracks into the side of an LRV going in the same direction.

For an emergency to be properly tested, the LRV should have a range of passengers, including teenagers, infants and small children with parent, elderly people, and people with disabilities, including wheelchair bound. Coordination between light rail and emergency service staff will be important in the safe evacuation and medical treatment of injured passengers and operators. Based on experience in other rail systems, emergencies cannot be managed in an ad hoc way. The training of staff in the safe and courteous handling of passengers is an important foundation for a safe and successful long-term operation.

Thankfully, rail crashes are very rare, but the consequences can be severe. In the last year, worldwide there have been only two light rail crashes involving injuries—11 in Melbourne and 35 in Rotterdam, but of course they make headline news. Nevertheless, having a fully trained staff will first avoid accidents and second mitigate the effects when they do occur.

4.9.5 Depot

The depot serves a number of functions, including location and LRV daily cleaning, overnight stabling, routine maintenance, accident repairs, daily staff briefings and training, system control, and security. With the infrastructure fixed, commissioning the depot will only show how the fixed assets can be managed better and the staff trained to operate the system efficiently. Over time, staff will find shortcuts and attempt to abuse the facilities. The role of commissioning is to try to identify how this might arise and put into place procedures to ensure that staff perform their functions safely and efficiently.

Commissioning the depot will include testing all the equipment, including hoists and jacks for lifting LRVs. If the control center is in the depot, all its functions, including alarms, switches, communications equipment, radio, etc., will need to be proven. Because an emergency exercise will be coordinated from the control center, liaison with emergency services should include inspection tours and meeting the control center team. Safety features like drains, ventilation, and ductwork also will also to be commissioned. Finally, all health and safety signage and facilities need to be checked and the depot staff fully trained.

4.9.6 Service

In commissioning the system, there will be a period of running the service without passengers so that operating staff can become familiar with the equipment and its operating parameters. This period will be between a week and a month. This allows timetables to be fine-tuned, especially for any peak/off-peak adjustments. At the peak, with higher volumes of passengers, boarding and alighting will take longer, so stop times will increase. This might hold up a subsequent LRV or mean that it misses its allotted time at a junction.

The system commissioning also will include two-way radios that will allow the operation to be managed in real time when problems arise. It also will allow staff rostering, including crew relief for meal breaks and end of shift, to be monitored. Smooth changeovers are important for running a reliable service. A terminal station is the best place for staff to change shifts, since the vehicles will be scheduled to stay there for a few minutes and any delay in staff changeover will have a minimal effect on service reliability.

When the Sheffield Supertramway was being commissioned during November 1995, a sudden snowstorm halted all traffic in the city except the trams, which ran empty. That certainly sent an important message about the reliability of LRVs in inclement weather, as frustrated travelers stuck in snowbound cars would have been happy to take an LRV.

4.9.7 Out of Service and Emergencies

No light rail system can expect to operate without occasional out-of-service problems, like a rail vehicle breaking down or a road vehicle blocking the tracks. Infrequently, there also will be emergencies, some of which will occur on the tracks (e.g., an LRV or other road vehicle crash). Some emergencies will be off-track, like a fire in a property adjacent to the tracks, when the fire department will require the OHL power to be turned off to allow firefighters safe access to the scene. This also will mean that the service has to be suspended.

Experience has shown that it is not realistic to expect the control team to ad lib its way out of every situation. The reason for this is that, in normal service, the amount of information handled by the control room staff is minimal (Cuffe 1989). During an incident, the data received by the staff increases manyfold. Often, the wrong information is acted upon, which makes the situation worse. Part of the commissioning period should cover the training of the control team to handle incidents, both as training exercises and, as discussed in Section 4.9.4, in responding to simulated incidents with real people on-site (e.g., at a station or road junction). Such training needs to include the failure of communication equipment.

This allows the control team to be given constructive feedback on their performance, so that when a real incident occurs, the team will be ready to make the decisions that best liaise with emergency services and maximize the safety of passengers on the system and others nearby. Strategies can be developed for restoring normal service more quickly. Such training exercises should be repeated regularly to maintain efficiency and confidence.

4.10 Maintenance and Repairs

4.10.1 Track

The two most common track problems are wear and corrugations. Track wear will be most noticeable around sharp curves (under 30-m radius) (see Figure 4.5), where there will be flange grinding into the rail side of the outer rail and into the groove keeper of the inner rail. Flange lubrication is a way to reduce this wear, as well as the wheel squeal which is associated with it. Similarly, reducing vehicle operating speed around curves also will reduce this wear. Initially, rail grinding will restore the railhead profile and therefore reduce the risk of derailments due to the flanges of the wheels climbing worn rail sides. Ultimately, even reprofiling will not fix the problem because the track will be out of gauge tolerance and the groove keeper will be worn through or broken off, creating another derailment source (Figure 4.47). This premature wear of a curved rail keeper led to the early replacement of a curve in 2007 on the Croydon tramway near East Croydon Station.

Railhead corrugations are worse when rails are laid directly on rigid bearers, whether concrete pads or sleepers. Corrugations are a series of hard ridges across the railhead with a regular wavelength. Tram rail corrugations usually have a wavelength of about 100 mm or less. These are due to hunting of the wheel sets when bogies oscillate uncontrollably across the track that work hardens from wheel impacts into zones about 2 mm wide and deep and allows the wheels to abrade the softer head between hard spots.

Unless caught early, the hard zones will propagate vertically right through the railhead. The only remedial treatment is railhead grinding to below the zone of hardness, usually more than 2 mm in depth. If the corrugations are not ground out soon enough, the hard spots will propagate right through the railhead. Rail grinding will not eliminate that problem. Only rail replacement can tackle it, and this is a high-cost repair because otherwise serviceable rails have to be scrapped.

Failures of crossings and supports (Figure 4.48) are a sign of either poor design or inadequate maintenance. The example in Figure 4.48 from Sacra-

Figure 4.47 Croydon broken rail keeper (Photo: L. Lesley)

mento shows a railroad design crossing used for a light rail system. The paving blocks are in poor condition and show that LRV wheels have been running over the blocks. The most likely cause of this failure is fatigue related, where the wheels drop into the crossing gap, hammering the foundation and disturbing the paving blocks. On a railroad, this is not a problem, as the crossing is supported on an "elastic" tie (sleeper) and there is no paving.

For segregated tracks, monitoring the condition of ballast and drainage will be important to maintaining a safe and comfortable ride in LRVs. Where grass has been planted around the tracks, the passage of LRVs will keep the grass mowed, but outside the LRV swept path, grass will need to be mowed regularly during the growing season. Alternately, green slow-growing plants that need less maintenance (e.g., sedums) could be used.

Regular maintenance tasks for tram tracks include keeping the grooves and drains clear of debris, so that the railhead will remain above standing water when it rains. Lubrication of switches is important, as is keeping tongues and blades clean of debris, so that they operate smoothly without danger of LRVs being derailed by not following the right path. Normally, switches are inspected weekly unless a driver reports a problem.

Where tracks are embedded in streets, debris is also deposited by other sources (e.g., motor vehicles, pedestrians, etc.). Track grooves can be cleaned out mechanically, for example with large vacuum cleaners. Switches require manual attention weekly at busy places and at least monthly elsewhere. In wintry weather, where tram tracks do not share salted roads, the operator will

Figure 4.48 Crossing failure in Sacramento (Photo: L. Lesley)

have to plow or sweep snow from the tracks (Figure 4.49) and defrost switch blade pivots and blade throw slots. Some systems also operate LRVs overnight to keep tracks clear for the start of passenger service in the morning.

The operator also will have to observe the condition of the highway pavement around tram tracks, since any delamination between the highway pavement and rails is a source of water getting into the track foundation. If the pavement is shared with heavy road vehicles, water pumped into cracks by road vehicle tires can lead to track failure, as has happened several times with the shared tracks on Moseley Street in Manchester. The delamination of road pavement from tram rails can be caused by both road vehicles and railcars, but the light rail system operator is the one that will suffer the damage and therefore has an incentive to rectify this as quickly as possible.

Most track maintenance and many repairs can be undertaken without having to suspend service, although road traffic may have to be temporarily

Figure 4.49 Snow-blocked light rail track (Photo: L. Lesley)

diverted from the work site. Special maintenance vehicles equipped with power tools have made track maintenance much faster and less costly.

4.10.2 Overhead Lines and Power Supply

OHLs with carbon contact strip pantographs should last about 30 years before wear reduces the mechanical strength or increases the electrical resistance to a point where replacement is needed. Normally, OHLs properly installed should be nearly maintenance free, but will need repair if damaged by a broken pantograph or hit by a high road vehicle (e.g., a crane boom on a transporter). In this case, the OHL should be repaired to the original profile, depending on the degree of damage.

Traditionally, a rail tower wagon has been used to gain access to an OHL. Today, a standard road self-propelled mobile "cherry-picker" work platform is the most economical approach as well as the most flexible. Such a vehicle can be fitted with flanged dolly wheels for use on unpaved off-street tracks. Maintenance or repair of OHLs also can be due to adverse weather conditions, wind damage, or thermal displacement. Similarly, if an OHL is suspended from adjacent buildings, demolition will require planting a (temporary) pole to support it.

Regular checking of the OHL supply cables is necessary to ensure that the insulation has not deteriorated, joints have not failed, or some mechanical damage has not occurred. Similarly, the feeder box isolators and circuit breakers in the substations need regular (monthly) checking to ensure that they are working in both the "off" and "on" positions. In particular, checking must ensure that there are no short circuits.

4.10.3 Vehicles

The manufacturers of tramcars should supply a maintenance schedule and manual. Two of the biggest areas for maintenance or repair are the wheel profiles and damage due to passenger use.

Light rail wheel profiles can be damaged by grinding rails, sideways into flanges, from wheels locking during braking and wearing a flat when the wheel is not truly circular, or from spinning on acceleration and therefore damaging the whole tire metallurgy by overheating. The normal method of recovering damaged wheels is by the use of a lathe and reprofiling the tire (and/or flange) by turning back to an acceptable profile.

Some systems will have underfloor lathes in the depot and can turn a pair of wheels on the same axle at the same time. In others, the whole axle/wheel set has to be removed and put into a floor lathe. Finally, some operators of vehicles (e.g., PCC, City Class, etc.) can remove the wheels individually (Figure 4.50) and re-turn them on a bench lathe.

Turning the wheels will remove metal from the tire and at least shorten its life. Reducing wheel damage in normal operation is possible by the use of flange lubricators, which also will reduce rail damage on curves, and having slip/slide (spin/skid) protection in the traction control system. This prevents wheels from locking during braking and in acceleration keeps the wheels within the maximum track adhesion available without spinning/slipping. Keeping the wheel profile true not only reduces maintenance because there is less damage from vibration but also improves the ride quality.

Passengers damage rail vehicles accidentally during use and, sadly, too often by vandalism. Dirty clothes, seat abrasion, and damage to floor surfaces from shoes are the most common causes. Keeping LRVs clean and looking smart is no different than maintaining hotels, hospitals, or any other public facilities. There is evidence which shows that vehicles kept clean will not be damaged as much by passengers. Indeed, operators can enlist passenger support in keeping the system presentable through social pressures (e.g., "do not litter" or "keep clean" signs).

Vandalism, on the other hand, is more difficult to address, and the usual remedy of making surfaces more vandal resistant usually merely deflects attention to other areas. Window scratching and graffiti are two types of

Figure 4.50 Individually removable City Class wheel (PCC inset) (Photo: L. Lesley)

vandalism that are expensive to fix and have the greatest impact on regular passengers. The extreme repair is glass replacement, which is expensive. For shallow scratches, acrylic-based fillers are an option. Installation of closed-circuit televisions on vehicles can provide evidence to prosecute vandals. A more positive and cheaper approach is to deter vandalism so as to avoid the need for repair or replacement. Here, applied psychology can be effective, including supervision by staff and other passengers and programs with schools and youth groups to channel these destructive energies to more creative avenues. On some systems, playing classical music deterred young people from congregating, which can be unnerving for other passengers. Music also can be used to create a calm riding atmosphere.

Window scratching is the most costly damage due to vandalism and annoys passengers (Figure 4.51). Vinyl films can be applied to the inside surfaces of windows, so that if a window is scratched, the film can be removed and replaced without having to take the glass out. Acrylic-based compounds also can be applied to scratches to reduce their visibility or remove them. A good customer care program is based on motivated operating staff, fully trained to avoid conflicts with passengers and willing to take firm action before passen-

Figure 4.51 Scratched door glass and graffiti in depot (Photo: L. Lesley)

ger disruption turns into physical damage of the vehicle or injury to other passengers.

4.10.4 Stations and Other Infrastructure

Light rail stops and stations need regular cleaning, and it may make sense where there is street cleaning to contract the local municipality to clean station platforms at the same time as the rest of the street. For off-street lines, dedicated cleaning teams are needed on a daily basis to collect litter and to repair damage. Smashing shelter glass at secluded stops can be a temptation for vandals. Stations designed on the principles of defensible space are less likely to be vandalized.

A rapid repair strategy is the most effective way to reduce such damage. This was demonstrated in Bradford, where the time it took to replace broken glass in a bus shelter was shortened from within a week to within 24 hours. This virtually stopped shelter damage within a month. British Telecom had nearly the same experience with telephone booth damage.

The lesson from these experiences is that preemptive maintenance is the best strategy to avoid expensive repairs. Having a pleasant waiting environment is the best way to build passenger confidence in the system and, with that patronage and revenue, to fund the service. A well-used station also has the benefit of social pressure reducing or avoiding vandalism or graffiti. Closed-circuit television cannot replace the presence of people, but it can provide evidence for prosecution, if the system is working and the camera is pointing in the right direction.

4.11 Refurbishment and Enhancement

All light rail systems require regular updating and refurbishment, both to replace physically worn-out assets as well as provide a more contemporary look for stations and vehicles. This should be included in the business plan and can be funded by regular depreciation in the annual accounts or by offsetting an operating deficit in one year with profits from operating surpluses over following years. Whichever way refurbishment and enhancement are funded, if they are not in the budget, then the system cannot maintain long-term viability.

Refurbishment, however it is funded, needs to be carefully planned and should involve existing passengers and also targeted passengers, to select styles and colors that will win the greatest approval. Refurbishment must have engineering inputs, as well as design and styling inputs. Replacement of equipment usually will be on the basis of an economic appraisal. Will it be less expensive overall to purchase a new item than to continue to maintain the old item? This evaluation also will include taking into account improved performance and the ability to satisfy more passenger expectations. As advances in equipment development continue to be made, it is very likely that the replacement equipment will have greater functionality and be less costly to operate and maintain.

4.11.1 Vehicles

Passengers always compare public transport vehicles and light rail cars with contemporary private motor cars. It is impossible for railcars to have every gizmo found in modern motor cars. In terms of presentation, style, comfort, and convenience, however, LRVs should be every bit as good, and with appropriate industrial designers, there is no reason why their interior appearance cannot be superior to automobiles.

Another reason for refurbishment is changes in legislation that require additional comfort or safety features for passengers or staff. An example of this is the requirement to make transit vehicles accessible for people with disabilities. Retrofitting facilities usually means compromises or reducing comfort or capacity for other passengers. Similarly, providing real-time information for passengers inside the LRV as it progresses along the route is not just a legal requirement in many countries but an important market require-

Figure 4.52 Supertram before refurbishment (Photo: L. Lesley)

ment. This is especially the case where attracting car commuters is critical, since passenger orientation is important to building passenger confidence. Car commuters are used to roads with good signage that directs them to their destination. For transit, the signage must be at least as good and the operation as reliable as clockwork.

Probably the easiest and least expensive way to change the look of an LRV is a new color scheme (Figure 4.52) and new seat coverings. Where seat cushions are worn, they should at the least be replaced. Better still, new seats could be installed, which would allow a new seating arrangement and further enhance both the look of the vehicle and the comfort of passengers.

During the operating life of the fleet, a record of reasons for delays and breakdowns is kept. This is an aid in determining the most frequent cause of such problems, which usually is vehicle door failure. During refurbishment, improvements to these critical parts can be implemented to reduce downtime during future operating periods and improve fleet reliability and therefore passenger satisfaction.

Clearly, such makeovers will require vehicles to be out of service. This has two consequences: first, the fleet available for normal service will be reduced, which may result in less frequent service; second, workshop space will be occupied, which is then not available for routine maintenance. An alternative in the second case is to subcontract the refurbishment to an outside company. This means that vehicles will leave the system on a rotation basis. Which of the two approaches is chosen will depend on the level of refurbishment and how accurately it can be scheduled.

The operator may, however, acquire some spare vehicles, by purchase or loan, to cover for those taken out of service for refurbishment. This is usually possible where neighboring cities have similar or compatible vehicles that could operate interchangeably.

4.11.2 Stations

Refurbishing stations may be due to changes in legislation, as discussed for vehicles, especially the need to make systems accessible for people with disabilities. Level and ramped access usually can be provided for light rail stops in streets, but the downside is that elderly passengers in particular can slip and fall on slopes in wet or icy weather, a common cause of fractured hips or legs.

On the other hand, making stations and vehicles accessible helps parents with young children in carriages or strollers, for whom the only alternative form of travel might be the family car. Indeed, it was the unexpected off-peak use of the Manchester Metrolink by mostly women with small children from one-car households that enabled the first line to reach early operating profitability. It also allowed a higher frequency of service throughout the day, by a frequent "turn up and go" service, without having to know the timetable. This generated further trips in other sectors of the population.

Closing stations for refurbishment also will be challenging (Figure 4.53), since the busiest stations will be the ones that experience the most wear and tear. Undertaking refurbishment at such stations while still in use will require a time-critical program, including overnight work when no passenger service runs. Similar to vehicles, changing the décor of stations is the fastest way to

Figure 4.53 Milan street station—candidate for refurbishment? (Photo: E. Bostock)

improve the passenger waiting environment, and the refurbishment also can include better information and a public address system.

Anecdotally, young people meeting and loitering at LRV stations can be disconcerting to many passengers. The Tyne and Wear Metro addressed this by using a public address system to broadcast classical music, which had the effect of moving groups of young people away from stations and perhaps making passengers who like classical music happy.

Where rail stations are on street-tracked sections, any refurbishment must also consider the local environment, including other street furniture, fronts of buildings, etc. This also may provide an opportunity for a themed refurbishment sponsored by local businesses.

A light rail operator will try to run a frequent and reliable service so that passenger waiting time is short. Nevertheless, during a heat wave, rain, or strong wind, some shelter will be important, no matter how short the wait. How much will depend on usage. The type of shelter will depend on budget, corporate identity, design innovation, and prevailing weather conditions.

There are certainly plenty of box-like "bus shelters" on the market which can be bought "off the shelf." A progressive operator will, however, try to offer an upscale image and therefore will want to provide shelter that has a distinctive flair. Giving the rail service a unique identity is important in creating and enhancing the brand image. A light rail system should be iconic in its city. At a time when multinational brands festoon city streets worldwide, a light rail system can be one, if not the most important, civic identifier.

Good transit design creates a loyal patronage base. The 1930s schematic London Underground Map is now copied throughout the world because it simplifies the transit connections, making it understandable to passengers who cannot read a geographical map. Building a local identity will be important to a strong brand locally. The design of stations is an important part of this.

4.11.3 Other Infrastructure Works

Other aspects of a light rail system that will benefit from regular maintenance or updating include the traction poles, equipment boxes in public places, and the public face of the depot. In addition to the stations, the traction poles and OHLs are always in the public eye. Unless a motor vehicle collides with a traction pole, there are normally few reasons to pay attention to this aspect of the system. Regular painting, however, can have a big impact on the urban streetscape and if integrated with the station house style will be an important marketing tool for building and retaining patronage for the tramway.

Where climate permits, hanging baskets of flowers can be put on traction poles to improve the city ambiance. Local residents can be recruited to water

Figure 4.54 Tidy depot in Berlin (Photo: BVG)

the flowers regularly. Similarly, traction poles can be used to fly flags and banners as part of a local carnival or festival. While such measures are not directly required for light rail operation, they do help to establish the operator as part of the local community and therefore within the collective consciousness.

Depots are seen mostly by those who live near them. It is important that the depot is a good neighbor, presenting a tidy and clean image (Figure 4.54). Like any other engineering facility, obsolete equipment can accumulate; therefore, regular tidying up and clearing out will not only help maintain a good image but also improve working efficiency. Better staff morale can be fostered by a policy of good housekeeping in the depot. Working in an organized and tidy facility improves staff productivity and reduces the risk of accidents. This is a variation of "lean manufacturing," for which there is now a large supporting body of literature.

Very rarely can a complete light rail system be built all at once, if for no other reason than the availability of capital to fund construction and equipment purchases. Building a system in stages is, therefore, the norm. This also means that patronage and therefore revenue will increase as each stage is opened. Generating revenue early will give funders confidence for further investments. Confirming the projected figures for the patronage of an extension is critical in all business plans, to guarantee that future investments will be financially viable. This was an area of considerable criticism in the April 2004 UK National Audit Office report on light rail projects. Refining patronage forecasts based on actual usage means that the economic evaluation of future lines can be made with considerably more confidence than the initial line.

In building the first line, there will be obvious places where future route extensions can begin. The question for the design engineer is whether to lay the junctions in anticipation of future extensions or to disrupt the operating line when future extensions are built. In an urban center, usually it makes sense to install junctions in anticipation of extensions, whereas installing new junctions in the suburbs causes less disruption and can be done as needed. In any case, even over a 10-year period, which is not unusual for building a full network, priorities will change, and new urban development may justify a line before existing areas, especially if the developer is willing to contribute to the cost.

4.12 System Extension and Development

4.12.1 Increase Route Capacity

Before new lines are built, it often makes considerable sense to see how much extra traffic can be attracted to the existing line(s). If this can be done with operational or management measures that need only limited capital investments, all the better. There are two simple ways to increase capacity within the existing capital and infrastructure assets: either increase the capacity of rail vehicles or increase the number of vehicles per hour.

The first often can be achieved by replacing seats with standing space. This adds more effective standing space in peak periods. Replacing 4 seats with standing space will create about 15 standing spaces, a net increase of 11, which in a 200-passenger vehicle is about 6%.

A compromise is the use of seats that can be folded back. Regular seating would be used during off-peak times, whereas the equivalent standing space would be used during peak hours. The extra space also would provide room for wheelchairs or baby strollers/carriages. Passenger surveys should be conducted to determine the most acceptable way in which this can be achieved.

The second method is to increase the number of vehicles per consist per hour. Many light rail systems use two or three coupled cars in a consist (Figure 4.55). Running longer consists, of course, means having longer platforms at stops and modifying traffic management and signal settings to allow safe crossing at busy junctions. Both will require some investment, but less than a new line to achieve the same patronage increase.

Alternatively, more consists per hour can be scheduled. If the vehicle fleet is fixed, then either the running times have to be reduced or the service recast, so that there is a more intense service over the busiest part of the route.

With operational experience, drivers can become more competent and confident; the timetable can therefore be tightened up to exploit the improved efficiency. Thus, more runs per day can be achieved, increasing both vehicle and staff productivity. Further discussion on this can be found in Section 4.12.4. Suffice it to say that there is always scope for improving

Figure 4.55 Long platforms for light rail two-car consists in Dusseldorf (Photo: City Transport Information)

performance, and every extra journey squeezed from the existing fleet reduces unit operating costs and increases revenue, creating a positive impact on the annual financial returns.

The next stage is to increase the effective use of the fleet, either by running more frequently or by coupling vehicles to make consist trains. This may mean that the station platforms have to be extended. In the earlier example, a 6-minute service required a fleet of 10 vehicles. If the round-trip time cannot be reduced, a 5-minute service can be provided by using 2 extra vehicles, increasing the fleet to 12 (20% increase). The most frequent urban light rail service historically was every 60 seconds, but with other road traffic, the practical limit is now probably about every 2 minutes. There is still considerable scope to increase frequency, capacity, and hence patronage on an existing line.

Reducing layover time at the terminal can be an easy way to increase frequency. This, however, requires extra crews at the terminal. The incoming crew can take a break and its vehicle would be taken out by a relief crew that brought in an earlier service. This stepping back of crews in taking out a later LRV requires good staff rostering and supervision, since a replacement crew that arrives late will severely disrupt service.

4.12.2 New Routes

New routes offer the tramway operator both a challenge and an opportunity. If there is only one rail line, adding a second geometrically increases the journey opportunities (with interchange) available. This is network synergy. In building a second line, and then a third, lessons will have been learned from the first in such areas as station design, track layout, and other operational aspects that can be incorporated in the new line. The object is to either save costs or improve passenger comfort or service. Better still, both can be addressed.

The three main financial aspects of a tramway—capital costs, operating costs, and revenue—will have been confirmed and the financial models used recalibrated. This will give even better levels of prediction. Therefore, the light rail operator can promote a second line with considerably more confidence than the first. This also will mean that funders should be willing to offer better terms, since they too will be more confident of an attractive return on their investment.

Further, the public and highway authorities will better understand what a light rail system requires. Having seen the performance of the first line, especially if significant car trips are attracted and traffic is reduced, they usually will be willing to offer more traffic priority over other vehicles. The operator also can plan for more convenient locations for stops.

Indeed, there may even be private developers willing to fund a new station integrated with this property development, as occurred at Ludwigshafen (see Figure 3.9), where a department store provided a new rail stop within the store. This made it very convenient for passengers to shop there. Other possible areas for similar cooperation are railway stations and airports, where the light rail can feed or distribute traffic to and from those hubs, allowing passenger volume growth without having to add to automobile traffic or add more parking spaces.

Usually the biggest operational question is whether the rail vehicles on the new line can use the existing depot or a new depot (site) is needed. In the latter case, the new depot might only provide stabling and vehicle cleaning and valeting facilities, with trams going to the original depot for maintenance or repair, thus creating economies of scale and therefore unit cost savings. This also would give the operator the opportunity to build a state-of-the-art maintenance facility and demote the original depot to purely overnight stabling.

4.12.3 New Service Pattern

A second (or third) line will provide the opportunity to introduce a new service pattern, building on the experience of the first line. There are basically two route strategies: many to many at infrequent intervals or few to few at more frequent intervals. Market research on passenger behavior shows high sensitivity to (perceived) service frequency and hence waiting times.

Figure 4.56 Light rail/trolley coach interchange station in Zurich (Photo: City Transport Information)

A convenient rule of thumb is that the service frequency, and hence passenger waiting time, should be less than the in-vehicle time if service is to attract passengers (Cudahy 1995). A light rail line with an outer-terminal-to-city-center journey time of 15 minutes will attract few passengers if the service frequency is every 20 minutes. Ideally, the service frequency should be about 30% of the in-vehicle journey time. In this example, patronage would be maximized at a frequency of every 5 minutes or better.

If the operator is planning high-frequency service, this will dictate a few-to-few route pattern. Maximizing the journey opportunities from such frequent service means having dedicated interchange station(s). This was done in Zurich (Figure 4.56), where connecting timings allow passengers to alight from one vehicle and almost immediately board the connecting service. At some interchange stations, like Zurich, passengers can buy a drink, something to eat, or a newspaper.

Even if this level of timetable reliability is not possible, with a high frequency of service (e.g., every 5 minutes) the maximum interchange time will be under 5 minutes and the average time about 2.5 minutes. Drivers would be trained to ensure that, where possible, connections are made. There is nothing more frustrating for interchange passengers than watching their connecting vehicle pull away, when they could have made the connection if the vehicle had waited a few seconds. Therefore, the design and location of new interchange stations are critical in building a new line. The interchange might be located where the two lines intersect, usually in or near the central area, or if the new line branches off the existing line, it could be at the junction station.

4.12.4 Intensifying Rolling Stock Performance

Once a new system is running and the drivers have become confident in operating the service, there are three ways to speed up the timetable: faster running speed, shorter stop times, and shorter layovers at terminals. This is important if costs are to be kept low by improved productivity and therefore more journeys per hour from each vehicle or consist. When LRVs are stopped, costs are incurred and little revenue is generated. Spending a lot of time at rail stops to allow passengers to get off or board reduces the efficiency of the system. Drivers can use the public address system to encourage passengers to use all available doors and therefore reduce station time. On- or off-board ticket machines can be used to reduce or eliminate purchases from the driver and an honor system combined with random ticket checking can significantly reduce stop time for little capital investment.

Traffic signals and traffic congestion also delay light rail service. In cooperation with the local highway authority, measures to give LRVs priority over other traffic will reduce running time and therefore make vehicles more productive. Within the parameters of existing speed limits, faster running times may be achieved by faster rates of acceleration and braking, which together with the above methods also can reduce running time. The maximum rates of acceleration/braking will depend on the traction equipment and the passengers. The highest acceleration that standing passengers can comfortably withstand is 6 ft/s^2 (1.8 m/s^2). Typical European LRVs are set for 1.2 m/s^2 (4 ft/s^2).

Finally, time at the terminal also can be reduced by stepping crews back, leaving only enough time for crews to switch vehicles. This means that the LRV can return immediately, while the crew takes a break. This requires at least one more crew than the number of vehicles on the route and two more if the same is done at the other terminus.

As an example, if the existing round-trip time is 60 minutes with a tram every 6 minutes, reducing the round-trip time to 50 minutes means that service every 5 minutes can be offered with the existing fleet. Computer models can be used to confirm that the new running time is practical within the normal variability and probability of more passengers or traffic delays.

Indeed, in the initial phase where the patronage generated by 6-minute service is now carried by 5-minute service, station time can be reduced further since fewer passengers (about 17% less) will be boarding or alighting, on average, each time an LRV stops. This might even be reduced to the minimum needed to open and close doors, further speeding up service. On the other hand, more frequent and slightly faster service will generate more patronage, which might slightly increase station time back to the original, but with 20% more passengers carried.

There are other ways to improve rolling stock performance. Normally, the morning and evening peaks require the full fleet to be available. Rolling stock has to be maintained. If minor maintenance can be scheduled for the interpeak period or overnight, then the full fleet can be used during the peak and some vehicles taken out of service during the off-peak for routine maintenance.

4.13 Engineering for Freight on Light Rail

For light rail systems that share existing railroad tracks with freight trains, engineering will be required to ensure that the light rail infrastructure does not impede freight train movement. This relates in particular to station platforms, as light rail cars will normally be narrower than rail wagons. One way to achieve this is to have light rail stations on a side loop off the main line.

Where a light rail system is to be adapted to carry freight, there may be obstacles, including street furniture such as lampposts that will have to be relocated. Contemporary LRVs are wider than maritime (ISO) containers, and therefore they can be carried on adapted LRVs, provided there is sufficient clearance on curves. If a 40-ft (12-m) container can sit on a modified LRV, which clears the curves, then there will be no problem.

For smaller consignments, the wheel secure cage/pallet is probably the most versatile unit. It can be wheeled on/off LRVs with level platform boarding. Given the use of Web-based freight delivery schedules, arrangements can be made to meet the consignment at the station, wheel it to the nearby designated location, and return empty units to the consolidation depot for refilling.

There are engineering issues for stations. Any ramps should be gentle, as a wheeled pallet of soft drinks can be quite heavy. Access/exit routes must be wide enough to allow passengers to walk to or from the platform without obstruction from wheeled cages being moved. The literature on warehouse and depot design includes dimensions, radii, and gradients to guide designers of new light rail stations.

5

Affordable Light Rail

Affordability has a number of interpretations. In the context here, the meaning relates to a system that can be built with available funding, whether public or private, and operating revenue exceeds the operating costs (Figure 5.1). Therefore, according to the first part of this analysis, an affordable line does not need an operating subsidy. For a publicly funded system, this means it can be operated with only the initial capital grants. This assumes that there is an allowance for asset purchases, repair, maintenance, and replacement in due course. In a commercially funded system, the operating profit also must service the capital debt.

Figure 5.1 Low-cost starter system in San Diego (Photo: San Diego Transit)

There are two different ways in which such systems are justified:

- Cost-benefit analysis for a publicly funded project, where the public and communal benefits (e.g., reduced travel time, fewer accidents, less air pollution, etc.) exceed the capital costs
- The internal rate of return satisfies the investors, where the operating profit is large enough to service the investment capital for privately funded projects and shareholders receive a fair dividend for taking the risk of making an investment

5.1 Capital Costs

The capital cost of light rail systems is an area of considerable criticism in the US and UK. In its 2004 report, the UK National Audit Office wondered why projects in England often overran their initial capital cost estimate and in real terms were 100% higher than equivalent new projects in continental Europe. For publicly funded projects, the promoter (usually a transport authority) can be seduced by the availability of "free" capital grants from the government to develop grandiose plans, allowing for all possible future options and "gold plating" the infrastructure or vehicles.

Bidders for publicly funded light rail projects also benefit from the budget normally being in the public domain. It cannot be a surprise that bids often equal the available funding. Critics using a supply-side analysis compare the cost of new light rail vehicles with the equivalent capacity of buses. The difference used to be that light rail vehicles were five times the cost of buses (Lesley 2005). Today, the difference is closer to ten times. Intrinsically, the two types of vehicles are equally complex, but of course there are more companies that make buses than rail cars, so the competitive market is less strong for rail cars.

Open bidding might be a sensible strategy if public money were unlimited. For the federal or state government with limited (capital) funds, a new light rail system will be weighed in the political calculus against the opportunity to deliver other benefits to electors, including tax cuts. In the UK, a new tramway costs the same as four new district hospitals or 30 new secondary schools. It is therefore not surprising that no new projects have been funded in the UK since 1999 (Nottingham) and ambitious projects in Leeds, Liverpool, and Portsmouth were rejected in 2006 and in London in 2008. Extensions of the Manchester Metrolink have only been half funded (Figure 5.2), and projects in London cannot expect any public funding before 2018.

The discussion on value for money in light rail goes back at least 20 years (Lesley 1987). There are two ways to reduce the capital cost of light rail systems. The first is to build only what is actually needed to get a system up

Figure 5.2 Salford Quays route (Photo: Greater Manchester Passenger Transport Executive)

and running, and the second is by financial engineering, especially by having a short construction period so that passenger revenue can be generated early to reduce the outstanding debt more quickly. "In Net Present Value terms, the long term counts for little when a discount rate of 7% is applied" (Anon. 1986).

The results are highly sensitive even to small changes in the discount rate used for cost-benefit analysis. A light rail vehicle costs ten times as much as a bus. Purchasing such vehicles with money that costs 15% is nonsense and was difficult to justify even in 2009 when bank rates were about 1% per year. Governments often distribute capital grants over a number of years, which means that projects that could be built in a year often take up to four, ensuring that the pain of disruption due to construction is maximized.

Achieving affordable capital costs requires clear project objectives. This is another area of criticism by the UK National Audit Office 2004 report. Public authorities are tempted to think that new light rail systems can magically regenerate an economically deprived area or reduce social exclusion—which confuses cause and effect. The goal of a new transport system should be to divert existing car traffic flows, especially because these cars cause congestion and pollution. Attracting enough passengers to pay for the operating cost is another challenge for system promoters, especially if fares are set for political reasons rather than with reference to the local travel market.

The most direct rail route not only is likely to have the lowest capital cost, but by providing the shortest ride time also would be the most attractive and therefore divert the maximum number of passengers from cars. An example of this is the difference between the first Metrolink line in Manchester (between Altrincham and Bury) with the second between Cornbrook and Eccles.

The first line is direct, fast, and generates an operating profit of some £5 million per year. The second meanders around the "brownfield" Salford Quays and is slower than buses that run on the street. Not surprisingly, it operates at an annual loss of about £10 million.

After a direct route, avoiding the need for tunnels and bridges should be high on the promoter's agenda. Not only are these expensive to build, but they impose operational constraints and are costly to maintain. Often, the structures are justified in avoiding road traffic problems. On most city streets where a light rail route will pass, every light rail vehicle will stop at every station to allow passengers to board and alight. On the same city street, more than 99% of car traffic will be nonstop. The logic therefore is that if a subway is needed to separate light rail from other traffic, the other traffic should be diverted to the subway and the light rail should operate at grade, at street level on the street, maximizing the convenience for passengers.

This, however, is where an enterprising promoter will face road traffic problems head-on, by designing traffic management measures that give priority to light rail vehicles through time-sharing traffic signal plans and space sharing with exclusive tram lanes. This is a far more economic option than a subway and makes more stations affordable and located conveniently for passengers. It will require more simulation modeling to persuade city highway engineers that light rail can operate without adding to congestion, but this is much less costly than building tunnels.

The third area to achieve an economical capital cost is in the stations. By not being in tunnels or on viaducts, they immediately will be more economic to build and allow flexibility in siting to locate closest to the main traffic generators and attractors. Stations should be built to handle only the vehicles that will operate when the system opens. At some stage, if coupled operation into two-car consists is planned, then all stations should be designed so that platforms can be extended. Extra accesses will have to be available for the larger passenger flows generated.

It also may make sense to build one or two of the busiest stations with longer platforms from the start, so that two light rail vehicles can occupy a platform at the same time and provide a model on which future platform extensions can be based. The patronage forecasts also will predict the likely daily flows at each station. This will help determine the most economical way to provide for these flows, especially the peak period flows, while maintaining system safety and corporate style at all stops.

Finally, designing and building the infrastructure as simply as possible will reduce costs and result in the fewest future maintenance problems. The Merseytram project, aborted by the British government in 2006, planned for the city center four delta double-track junctions. Using the above analysis, none should have been needed, thus saving about 10% of the capital cost and considerable disruption due to construction.

The depot is another major capital investment. Promoters should avoid the temptation to build for any ultimate system size. Providing only for the initial fleet in terms of stabling, cleaning, valeting, and maintenance is the most economic option, provided land can be reserved for any future extensions. Not burdening the initial system with the capital costs appropriate for one much larger also will help to make it affordable. A poor example is the planned Merseytram depot at Gillmoss; designed for 18 light rail vehicles, it would have been twice as big as the Merseyrail Kirkdale depot, which stables 50 three-car Class 507/8 train sets.

The capital cost of the depot can be further reduced if the vehicle supplier has a maintenance contract. The system promoter can then trade off the savings in capital expenditure on maintenance equipment against the extra cost of contracting out maintenance.

The above discussion pertains to systems that are publicly funded. Raising the capital for privately funded systems is very different. Historically, virtually all tramways and railroads were privately funded (Figure 5.3). Most were built by contractors that were paid by the promoting company not totally in cash but in a mixture of cash and shares. On completion of the project, the company was then offered on a stock exchange and the contractor sold the shares on the stock market; this early form of the "dot-com" option generated more money than a straight cash contract would have. This method of financing projects lasted about 80 years.

Today, contractors usually want to be paid in cash, so the promoting company has to raise all the capital needed up front. Normally, however, a contractor will be one of a consortium that is promoting the project and will

Figure 5.3 19th century share certificates

be required to put in some capital, in the form of either funding or staff and other resources. The contractor also will be expected to charge an "at-cost" price, in return for profit sharing.

Usually the bulk of the capital is raised from merchant banks. These banks, however, expect that the promoters will invest at least 10% of the capital first. Even merchant banks divide the capital among a syndicate of other banks, as was done in constructing the Channel Tunnel. For the more modest capital required for a new light rail system, merchant banks also advise on a share issue.

Finally, the promoting company will have raised the capital with a combination of loans and shares to construct the new light rail system. Ideally, the more shares there are, the easier it is to achieve profitability because the interest on loans will have first call on any operating surplus. Shareholders will be paid only when the company is profitable. Once the system is open and revenue comes in, dividends can be increased by refinancing the debt. Further shares can be issued to pay off some or all of the bank loans.

Banks will require the interest on loans to be paid even if the company is not yet profitable. This leads to the company having to borrow more money. The Channel Tunnel Company paid as much to the banks to fund the project as it did to the contractors that built it.

Raising most of the capital from shares means, however, keeping the price of a share up so that shareholders can realize their investments by selling to other investors. Again, the Channel Tunnel project provides an example, albeit extreme, of the importance of share price. Investors initially paid £5 per share; halfway through construction and 4 years before opening, the price of a share reached £18. Four years after opening, when it was clear that traffic projections were overly optimistic, the price of a share sank to £0.50.

At this point, it became clear that no more money could be raised from shareholders and that the bank debt could not be serviced from revenue. A number of refinancing deals were arranged whereby the banks exchanged their loans for shares; this diluted the importance of the original shareholders, who today in total own less than 10% of the company. The Channel Tunnel is an extreme example of a privately funded rail project.

Capital for a light rail system is much smaller in comparison and the risks are less. It should be fundable from investors in the local area, who will benefit either as users or in terms of trade. Local investors can be offered free travel as an extra reward, on top of dividends.

The top priority of such private projects is maintaining confidence. This can be achieved by using reputable contractors and, as important, having competent management with good public relations skills. Meeting published deadlines is critical to maintaining confidence. Once one line has been completed as expected and service and revenue start, then the promoters will be able to raise further private capital more easily.

Figure 5.4 Charles Yerkes (Photo: Chicago L)

Charles Yerkes (Figure 5.4) made a fortune from the North Chicago City Railway, creating a large transportation empire. He then built up the Underground Group in the early 20th century in London. This group issued new shares for each new line built. The bulk of the underground system in London was constructed by this means. Similarly, the first generation of streetcar and tram systems was built by private funding until the Wall Street crash in 1929 made raising new funds from shares impossible. In England, British Electric Tractions Ltd. built nearly half the country's tramways. The company still exists, but its transport assets were nationalized in 1968 and today it is a general service company.

Some new light rail systems do not own their own fleets; instead, they lease the vehicles from a bank or other financial institution. This can be more economical than buying the fleet outright and certainly reduces the up-front capital investment, but the situation depends on the local taxation rules and credits. In those countries where there are tax advantages for large financial institutions to offset the lease income against the capital cost, then leasing will be an attractive option.

A lease and maintain deal can be part of the financing of a new system. A lease deal can be paid for on the basis of the availability of vehicles. The operator can fine-tune the revenue generated to fully meet the cost of rolling stock.

The disadvantage of a leasing arrangement is that the lessor will always want to have a fallback in case of default that allows the vehicles to be used on another system. This restricts the operator in terms of local customization of the trams. On the other hand, a vehicle lease deal that includes midlife refurbishment, with a rental review, can be an equally attractive way to operate and modernize the fleet without capital investment.

The lease premiums will appear as an operating cost and can be offset against any company tax payable at the end of the accounting year, which also will accelerate the date when the system is genuinely profitable. This shows the importance of financial engineering in minimizing (servicing) the capital debt and therefore the time when shareholders can be rewarded for their investment risk.

5.2 Operating Costs

Operating costs include all costs of consumables and noncapital items needed to operate the service. These costs include power consumption, staff wages and salaries, rent, leases, insurance, maintenance, spare parts, other materials, and cleaning. Operating costs can be carefully preplanned in advance of operations, and the commissioning period will allow budgets to be further refined. Normally, the biggest operating cost item is staff wages and salaries.

There is a temptation to make new light rail operations overly complex, which is one of the reasons why new systems have become uneconomic and were heavily criticized by the UK National Audit Office. In 1999, Joan Dunbrack of the Boston Transit Authority (MBTA) was visiting the Supertram system in Sheffield. When she toured the control room, the three staff members on duty who were controlling vehicle dispatching, power distribution, and the trackage explained how difficult it was to control a peak fleet of 25 trams. Ms. Dunbrack then said that she did the same job as the three controllers single-handed in Boston. She had a morning peak fleet of 250 trams to manage. Keeping operations simple reduces not only the operating costs but also the likelihood of things going (badly) wrong (Cuffe 1989). This also should reduce the capital costs.

Drivers (Figure 5.5), and conductors if they are used, will be the largest staff costs. In most systems, the morning and evening weekday peaks will need the largest staff turnout. Some operators use a combination of full-time staff to cover the base service operated throughout the day and part-time staff to cover the morning or evening peaks. Historically and in some systems, a hybrid split shift is used, whereby staff work both the morning and evening peak and have most of the morning and afternoon off. Staff share different shift patterns on a rotation basis. Agreements with unions also will have an impact on staff costs.

The hourly rate of pay, unless there is a salary contract, will reflect local labor market conditions. The staff that operate expensive machines in public environments where safety is critical probably will require a pay rate above the local norm for other service industries. The higher pay rate would attract higher caliber staff, unless there is a large pool of unemployed, in which case a lower pay rate could still attract and retain a highly motivated work force.

Figure 5.5 A Blackpool light rail operator (Photo: L. Lesley)

Rail systems find that staff are more stable and loyal and on average stay much longer than do bus drivers, who have a transferable skill.

The number of staff and the pay rate will determine the total staff costs. The norm for many light rail systems in the developed world is about three drivers/operating staff on the payroll for every vehicle in the operating fleet. This allows for coverage of an 18-hour or longer working day, holidays, sick days, and training requirements. In addition to pay, good management and nonpay incentives can have a large impact on improving and maintaining morale. One nonpay incentive is seniority in the choice of shifts, to encourage staff to stay, which reduces turnover and therefore recruitment and training costs.

The depot and workshop will be the next largest staff cost center. For an efficient operation, there should be enough work to keep all the staff fully occupied. For a newly opened system, rarely will there be significant maintenance, and unless drivers are careless, there will be little accident damage to repair. One option adopted by some systems is the use of multiskilled staff, so that when there are no workshop tasks, staff can be used for other duties, including operating the system. Indeed, there is much merit in having all staff, including the CEO, work an operating shift driving a light rail vehicle, perhaps once a month. In many systems, the depot staff also serve as the emergency team, assisting emergency services when an accident (e.g., rerail or rescue an overturned vehicle) happens.

While this has been a theoretical analysis, often light rail systems find it hard to succeed because of the perception that costs are higher than bus; real data, however, shows a different picture. Analysis of new light rail systems in US cities compared the operating and maintenance costs with bus service in the same places (Table 5.1). In most cases, the light rail transit has a lower

Table 5.1 Bus and light rail transit operating and maintenance costs in 2001

	Operating and maintenance cost ($ million)		*Passenger miles (million)*		*Cost ($) per passenger mile*	
City	*Bus*	*Light rail*	*Bus*	*Light rail*	*Bus*	*Light rail*
San Diego	58.2	26.5	153.5	152.7	0.38	0.17
St. Louis	98.8	19.2	150.4	95.9	0.66	0.20
Los Angeles	633.4	51.4	1332.2	170.5	0.48	0.30
Portland	120.9	22.3	237.0	63.3	0.51	0.35
Sacramento	41.8	14.8	79.5	39.4	0.53	0.38
Dallas	161.0	27.7	218.8	58.9	0.74	0.47
Baltimore	150.2	23.0	285.9	48.0	0.53	0.48
Denver	142.1	8.0	269.8	13.1	0.53	0.61
San Jose	145.9	26.2	202.4	33.0	0.72	0.79
			Weighted average		0.53	0.34

Data from www.lightrailnow.org/facts/fa_lrt02.htm

operating and maintenance cost per passenger mile than bus. The worst is Baltimore at 90% and the best is St. Louis at 30%, with a weighted average of 64%. These figures should give the light rail promoter confidence that by adopting the best, or nearly the best, practice, operating costs can be (much) lower than the equivalent bus costs and passenger revenue higher.

5.3 Patronage, Fares, and Revenue

This is the area of assessment that is most open to alternatives and interpretation, including overly optimistic forecasts, a major criticism of many recently opened light rail systems. "From where will the passengers come?" a transit executive once asked. The population living in the light rail hinterland will generate most of the patronage initially, and the revenue forecast must be based on this. These passengers will be attracted from other modes of transport, mainly private car, the mode that dominates trip making in most urban areas. To this, as appropriate, can be added park-and-ride from outside the catchment, diversion of trips from other destinations attracted by the convenience of the light rail system, and generation of trips that previously were not made within the area.

The number of trips made per week per person has stayed remarkably constant over the last 50 years, at about 21 in the UK. There is no evidence from elsewhere that trip making is significantly different or has grown over the past few decades. The reason for this is that going to work (or school)

Table 5.2 US commuter trip length

Year	Trip length (miles)
1983	9
1990	11
1995	12
2001	12

Source: Federal Highway Administration

Table 5.3 Netherlands commuter trip length

Year	Trip length (miles)
1985	5.4
2000	6.6

Table 5.4 Boulder, Colorado commuter trip length

Year	Trip length (miles)
1990	5.3
1994	6.2

is the dominant journey purpose and there are only so many workdays in a week.

What has changed is the mode of travel and trip length (Tables 5.2 to 5.4). The work journey more than doubled in length on average, and there was a major switch from public transport to private cars. In 2005, almost 70% of all work journeys in the UK were made by car, and only inner London had a significant use of public transport to get to work.

In the US, New York accounts for nearly 75% of all US rail transit trips, and most of those are commuters into Manhattan. In the rest of the US, metropolitan trips are dominated by cars; for example, more than 90% of all trips in Houston are by car. Less than 5% are by transit.

With concern about dependence on imported oil, environmental pollution, and traffic congestion, there will be a slow but growing modal transfer in the US from private car to transit and in particular light rail. Two reasons for this are the relative affordability of new light rail lines and their proven ability to attract car trips.

Of the 21 trips per person per week in the UK, how many can be attracted to light rail? There are well-established economic diversion curves that will predict the probability of a trip being made by one of several alternatives, including light rail. This is calculated on the basis of generalized costs. These bring together the facets that market research has shown influence choice of mode decisions. They include journey time, waiting and walking time, interchange time, comfort, convenience, cost, etc.

These facets have different impacts on the choice of mode decision and are therefore weighted to put them all in the same equivalent units: cost or more commonly time. The advantage of time as a constant measure is it is monetarily inflation proof. It is also more logical, since a shorter generalized time is correlated to more trips being made, just like the classic economic price/demand curve.

Obviously, part of the assessment is the number of trips that can be attracted to a (new) rail line and the market condition of alternative transport options, including the cost of car parking, bus frequency, taxi fares, etc. Indeed, any modeling of likely rail patronage also will include sensitivity analyses of how other modes of transport might react and therefore impact tramway patronage.

In two such studies, the impact of competitor bus operators charging zero fares was examined. In both cases, the result was a reduction in rail ridership by fewer than 25%. Zero bus fares, however, cannot be a long-term option for a bus operator. In many countries, it would be considered anticompetitive (e.g., illegal in the European Union). A bus operator that charges zero fares will either go bankrupt or will have to raise fares above the original level to replace the lost revenue. In either case, the rail operator will benefit.

The number of rail passengers is also sensitive to the fares charged. For politicians, low transit fares are a kind of mantra. However, little if any evidence exists to show that low fares attract significantly more passengers, including car users. This is because fares are usually less than 25% of the choice of mode decision making, and 75% is based on service quality. The low fare elasticity, usually between –0.3 and –0.4, confirms this.

In England, there was a South Yorkshire fare freeze during 1975 to 1986, the Merseyside low fares experiment between 1980 and 1986 (Hay 1986a, 1986b), the London "Fares Fair" plan between 1981 and 1982, and the current low fares regime in London. There was little statistically significant data to show that low fares attracted any trips from cars to bus, but many dependent bus users did make more bus trips. In the UK, the most prolific taxi users are in the lowest income quartile of households, which have the lowest level of car ownership. High fares are not a disincentive to taxi use, since cost is balanced against the quality of service obtained.

Therefore, a (new) rail system will want to position itself between the fare levels of bus and taxi. For a commercial system, this will be market driven. For a public system, if a political decision is made to have low fares, then there will have to be a subsidy or other compensation to make up for the lost revenue.

The principle for this has already been established in the European Union. In the UK, where senior citizens get free bus travel, the bus operators receive compensation on a "no-better/no-worse" basis. The operators receive the difference between the fares they would have collected from (fewer) senior citizens paying the full fare compared to what they actually get from seniors using free travel passes. Some operators complain that this is not enough. If

Table 5.5 Revenue sensitivity to rail fare in a new light rail system

Rail fare	*Patronage (million/year)*	*Revenue (£ million/year)*
0	6.0	0
£1	5.0	5.0
£2	4.5	9.0
£3	4.0	12.0

concessionary travel makes up a large part of bus patronage (37% on Greater Manchester in 2004–2005), more buses have to run more to provide extra capacity, which increases operating costs.

The number of passengers and the average fare charged together produce the total revenue. Again in a commercial project, the sensitivity of patronage and revenue to fare level was calculated (Table 5.5). The highest patronage (6 million passengers per year) was achieved with a zero rail fare. At an average fare of £1, 5 million passengers per year would be carried, generating revenue of £5 million a year. At a £2 average fare, 4.5 million passengers per year would generate £9 million a year, while an average fare of £3 would attract 4 million tram passengers and generate revenue of £12 million a year. In comparison, the equivalent average bus fare in the same town was £1.20 and taxi fare was £5.50.

In practice, a commercial rail operator would offer a variety of fares, perhaps copying the low-cost airlines, which maximize revenue yield from low fares for advance bookers and charge high fares for those who wait until the last minute or travel at peak times. Another cost model is that used by cell/mobile phone companies. Indeed, cell/mobile phones might be an efficient method to collect fares, since they are already being used for such diverse things as paying for parking and ring tones.

This discussion on revenue collection cannot be complete without considering passengers who travel without paying. No undertaking will ever reach 100% payment compliance. In European countries, many studies have been conducted to determine the unpaid portion of revenue. Typical figures are between 2 and 5%. In a supermarket, "shrinkage" from shoplifting can be as high as 10%. The rail operator will have to determine, provided this figure is static, what trying to collect the final 2% of revenue is worth.

One option is to install ticket-reading entry and exit gates at stations. This can only be practical on a closed system. Some systems have experimented with entry gates to the rail vehicles. In either case, a capital investment is required and maintenance costs are high. These can be reduced by using gates that are normally open and only close if an invalid ticket is presented. The rail operator must, however, weigh the extra revenue collected against making the system less attractive to "honest" passengers, who can easily travel in their cars.

Another revenue protection option is the use of inspectors, either traveling on vehicles or at busy stops, to check tickets as a deterrent against fraudulent travel. Many German rail systems use a sampling rate of 1 in 10 journeys and charge a "penalty" fare 10 times the normal or prepaid fare to those without valid tickets. In a new rail system, making it easy to pay the correct fare and the use of social pressure can be as effective as a revenue control inspector, which is another operating cost and can have a negative impact on patronage if perceived as the "gestapo" by the public.

Indeed, in 1995 the transport undertaking in Brussels was concerned about a slowly declining patronage and undertook market research to find out why people were not traveling by transit (bus and rail). One of the reasons given was the imposing appearance of the revenue inspectors, the "gestapo" as passengers nicknamed them. A clever public relations display included a new uniform that was less military in appearance and a citywide poster campaign. The posters had pictures of all the revenue inspectors with slogans like "Uncle John, here to help you get home safely." After 6 months, the image of Brussels transport had changed significantly for the better, and after 9 months patronage (and revenue) started to grow again.

An operator that forgets that the image of the rail system is as important as the hardware and training is in danger of having no passengers. In 1986, when bus services were deregulated in England, a new bus company (BEELINE BUZZ) set up in Manchester and used roadside posters along the planned route to promote a new service using minibuses. The established bus company, Greater Manchester Transport, reacted by running (expensive) advertisements on local television. Later, a follow-up public recognition survey was conducted. All respondents remembered the bus ads on television, but most thought they were on behalf of the BEELINE BUZZ Company. Image is both important and has to be developed and defended, but the choice of the wrong media can be an expensive mistake.

5.4 Financial Viability

At the very basic, financial viability means that passenger revenue should exceed operating costs for all light rail systems, since an operating surplus is needed to fund repairs and refurbishment. A public sector light rail system may set its fares below the market rate for political reasons, but it will then need a subsidy. There are many examples which show that this compensation is rarely adequate and operators have to make do and mend the service or let it become rundown (Cudahy 1995).

For a commercial light rail system, an operating surplus also will need to service the capital cost of building and equipping the system, whether by loan/debt, equity, or most likely a combination of the two. The operator, therefore, in the short term has to fine-tune the fares and revenue yield. In the medium term, the impact of patronage volume will affect fleet size, number of staff, etc., all of which can improve the profitability of operations. Even in public sector systems, fares may be raised in the peak period to encourage passengers to shift to an off-peak period, which would otherwise require additional vehicles. The extra revenue could be used to fund the purchase of more vehicles.

Monitoring costs and income should be a daily or, at most, weekly exercise. In some parts of the retail industry (e.g., food), stocking is updated daily and in such fast-moving sectors as fashion and clothing even weekly. The light rail operator must be able to take corrective action quickly if costs begin to rise or, more likely, revenue drops below the budget. Delaying action can make the eventual decisions expensive to introduce, often in a crisis situation, leading to a loss of staff morale and public credibility.

Financial "engineering" at this stage is as important as the physical engineering. Managing the cash flow, from which the ultimate operating surplus is derived, is a skilled task, especially as expenditures are uneven and the revenue flow is relatively smooth. Investing surpluses for later drawdown (e.g., for midlife refurbishments) could generate nearly half the total required from earned interest, assuming continuing relatively low levels of monetary inflation and interest rate stability.

Monitoring patronage will be an ongoing task for the traffic manager and, as in low-cost airlines, will require suitable strategies to cope with any downturns in demand. One option is joint ventures with important traffic generators to target people going to the cinema, shopping centers, etc. Market research can identify what is being done right and what needs to be improved. By definition, all organizations can operate better. Feedback from operating staff also should be solicited, since they are in daily contact with the customers who buy the tickets and pay their wages. The flip side to that is positive customer care, which by reputation will have an impact on patronage through personal recommendation to nonriding friends and neighbors.

5.5 Project Funding

5.5.1 Public Promoter

For a publicly funded project, there may be several sources of funding. Federal or state government can provide a grant that will be paid in installments against project progress. The government also can offer a low-interest loan, again paid against project progress, but the interest also will have to be serviced and the loan paid back from operating surpluses. Central government can give and underwrite local government credits for loans from commercial banks or international funders like the World Bank, International Monetary

Fund, European Investment Bank, etc. Again, the interest and capital will have to be repaid, but the advantage is a low interest rate because of the low level of risk associated with publicly promoted projects.

More recently, governments have used complex schemes like private finance initiatives or public-private partnerships (PPPs) to raise capital from private lenders. In both of these arrangements, the government transfers on paper the project risks to a private consortium of constructor, equipper, and operator. The consortium raises the investment capital to build and operate the system. The government gives the consortium a concession for a number of years, long enough for the original investment to be rewarded. Where an operating profit cannot be guaranteed, the government will offer an annual subvention to insulate the consortium.

The advantage of such schemes for the government is that the capital debt does not appear against the government's total debt, for purposes of fiscal policy or to satisfy external bodies like the International Monetary Fund that the government is not "living above its means." In practice, it is a sophisticated mortgage for which the government is the fallback and ultimately will have to repay the total cost. This was illustrated in the summer of 2007 when one of the London Tube PPP consortia went bankrupt and the project, with extra costs, had to be taken back by the public sector to ensure that the subway kept running and the modernization continued.

Private consortia also have developed strategies for increasing the profitability of such contracts. One way in which this is achieved is by refinancing the project when the infrastructure has been completed. The logic behind this is that the initial investment is very high risk with a high rate of interest from private lenders. Once the project has been completed, the infrastructure becomes the security against which to borrow at lower rates of interest.

This allows the original investment, which is usually put in by one or more members of the consortium, to be paid back. Members of the consortium thus make a trade profit from providing the service of building/equipping the system and then a financial profit by getting their capital investment back with interest. There are also examples where the new consortium then sells the concession to another investor, making another profit. In the end, such consortia can earn more than a 100% return on the original capital invested.

5.5.2 Private Promoter and Funding

Historically, most transport infrastructure has been promoted, funded, built, and operated by private companies. Many transport companies were subsequently taken into public ownership for a variety of reasons, including political ideology, national security, consumer and passenger protection, etc.

A private promoter that approaches a new tramway as a commercial venture is seeking to create a profitable business and to be able to pay back the initial capital costs. The private operator also will want to build up the

Figure 5.6 Commercial modernization (top) in London in 1931 and (bottom) in Brooklyn in 1934 (Top photo: London United Tramways Ltd.; bottom photo: US Info Service)

value of the asset, against which further investments can be generated (Figure 5.6). A private light rail promoter will therefore seek routes that generate the maximum revenue for a minimum capital cost. How will the funding for such enterprises be raised?

Historically, most private transport systems were built with funding in the form of shares. Many contractors were paid in shares. Once the system was built, the company was offered on the stock market. The original investors could then sell their shares to realize a good return for their investment. More recently, the dot-com investment boom of the early 21st century was funded in a similar way. Like rail in the 19th century, this was based on high-tech cutting-edge new companies formed to exploit the new opportunities of the Internet and e-trading.

A private promoter most likely will use loans and shares to fund a project. Indeed, any financial institution willing to lend for a new light rail project almost certainly will expect the promoter to put up at least 10% of the investment costs and take shares. In addition to merchant banks, international financial institutions like the World Bank, International Monetary Fund, and European Investment Bank will give loans at below high-risk commercial interest rates to help promote better "public" infrastructure and facilities. This has been one of the ways in which the privatized water companies in England have funded the massive modernization of water supply systems after years of underinvestment when in public ownership.

However a private promoter funds the building of the system, once it is open and revenue is being generated, other investors can assess the profitability of the project. The promoter can then offer the company on the stock market in an effort to attract further funds to expand and intensify the service. The advantage of shares as a source of investment capital is that there is no permanent debt servicing. Debt servicing reduces the profits, part of which have to be paid to shareholders as dividends. A small change in revenue, or operating costs, can mean a significant change in profitability and hence dividends.

The promise of dividends on shares, when a company is profitable, will be more attractive than interest on a loan. There is a second benefit of a capital gain from selling shares. Historically, company shares have appreciated in monetary value more than other commodities. Thus, even if a company pays little in the way of dividends, shareholders can be satisfied with the rise in value of their shares. Depending on the local tax structure, this might be financially more attractive to some investors. Even the fall in stock market values following the "credit crunch" in 2009 was short-lived and investor confidence returned by 2011, allowing new projects to be funded.

For a private promoter, the other major attraction of raising the needed capital by means of shares is that, unlike a loan, the capital costs do not have to be repaid. The shareholders can recover their original investment by selling their shares to other investors or to another company that wants to take over the rail system company. For these reasons, therefore, a private rail promoter will always try to maximize capital funding via shares and to keep the value of the shares high and growing, even if dividends are not always that high.

Table 5.6 10-year construction period and 7% interest per year

Year	*Costs*		*Interest*
	Infrastructure	*Vehicles*	
1	20	0	1.4
2	20	0	2.9
3	20	10	5.2
4	20	0	7.0
5	20	0	8.9
6	20	10	11.6
7	20	0	13.8
8	20	10	16.8
9	20	0	19.4
10	20	10	22.9
	Total cost = 350		

Table 5.7 5-year construction period and 7% interest per year

Year	*Costs*		*Interest*
	Infrastructure	*Vehicles*	
1	40	0	2.8
2	40	10	6.5
3	40	10	10.5
4	40	10	14.7
5	40	10	19.2
	Total cost = 294		

Table 5.6 shows that with interest, the total capital debt is £350 million, with an annual debt charge of £23 million. Table 5.7 shows that with interest, the total debt comes to £294 million, with an annual debt charge of £19 million.

A private promoter also will want to implement the project quickly so that revenue can be generated early to repay the debt more promptly and thus accrue less interest. The examples in Tables 5.7, 5.8, and 5.9 show the effect of shortening the construction period and also the impact of a change in the interest rate. Table 5.8 shows that with interest, the total capital debt is £311 million, with an annual debt charge of £26 million. Table 5.9 shows that with interest, the total capital debt comes to £266 million, with an annual debt charge of £17.4 million.

Comparing the examples in Tables 5.6 and 5.9 shows that shortening the construction period reduces the annual debt charge by about £6 million per

Table 5.8 5-year construction period and 9% interest per year

Year	Costs: Infrastructure	Costs: Vehicles	Interest
1	40	0	3.6
2	40	10	8.4
3	40	10	13.7
4	40	10	19.4
5	40	10	25.6
	Total cost = 311		

Table 5.9 2-year construction period and 7% interest per year

Year	Costs: Infrastructure	Costs: Vehicles	Interest
1	100	20	8.4
2	100	20	17.4
	Total cost = 265.8		

year, a 31% cost reduction. This gives the light rail operator either increased profitability or a longer period in which to build up patronage and hence revenue to reach the target operating profitability.

The examples in Tables 5.7 and 5.8 show how sensitive project debt is to changes of interest charges. Raising the interest rate from 7% to 9% increases the annual debt payment by £7 million per year, a 38% cost increase.

The examples in Tables 5.7 and 5.9 show that rapid construction in only 2 years reduces the annual debt charge from £19 million to £17.4 million (10%) and the overall debt from £294 million to £266 million.

A private light rail promoter will therefore try, as far as possible, to fund the project from shares rather than debt. If a loan has to be secured, then borrowing at a low interest rate and completing the project in the shortest possible time are the best options for achieving a financially viable project. On completion, the debt can be refinanced at a lower rate to reflect the asset created and lower associated risk. In the example in Table 5.8, if the project is refinanced at 5% per year, then the annual debt interest payment falls from £25.6 million to £11 million, which is £10 million less or a 64% saving on interest payments.

5.6 Economic Appraisal and Cost-Benefit Analysis

When investment capital is in short supply, as it normally is, an important economic concept is "the opportunity cost." If the capital were invested in another project, would it generate a higher return? Two methodologies have been developed to answer this question. The one used in the private sector is older and based on discounted cash flows over the life of the project to determine if there is a positive cash return—a profit. The second test is the internal rate of return. This cash accounting method has been adapted by public sector economists as cost-benefit analysis. It includes noncash items, like the environment, safety, travel time savings, etc.

5.6.1 Discounted Cash Flow

The accountant's shorthand for opportunity cost is the discount rate. This can be considered to be the cost of borrowing money. It is the way in which the private sector assesses how funds can be invested in capital projects. The logic behind discounted cash flow (DCF) is that there is an advantage to reaping benefits today and paying later. This is like a personal credit card.

The private sector derives less benefit from income later in the life of a project, when of course there is more uncertainty. Usually DCF is used for a 10-year projection period, since any benefits after that are likely to be overtaken by events. Such changes, as occurred in 2008–2009, include market volatility, credit drying up, a new competitor product, etc. This means that a new investment may be needed to retain the company's market share. If an investment project cannot be justified within 10 years, it is unlikely to be undertaken.

This immediately creates a problem for investments with long lives, such as light rail systems, power supply networks, generating capacity, etc. The DCF philosophy can still be applied—namely, the investment should pay for itself in 10 years. If it lasts for 30-plus years, then it will still be profitable but reduced by the need for increased maintenance.

A little digression here is useful. In West Germany in the 1980s, companies were examining the first wave of serious automation to improve labor productivity. Usually, such investments were undertaken even if the return was no better than the status quo. The reason was a tight labor market. Companies were paying staff increasing wages above the level of inflation. The argument was that as labor costs continued to grow faster than other costs, a capital investment would prove to be preferable and need less staffing (recruitment and training costs).

Table 5.10 Project DCF with 10-year construction period

	NPV cost		
Year	*Infrastructure*	*Vehicles*	*NPV revenue*
1	20	0	0
2	18.6	0	0
3	17.3	8.7	0
4	16.1	0	2.0
5	15.0	0	1.9
6	14.0	7.0	3.5
7	13.0	0	3.3
8	12.1	6.1	4.6
9	11.3	0	4.3
10	10.5	5.3	5.3
TOTAL	147.9	27.1	24.9

In DCF, annual income and expenditures are discounted to the present day by the discount rate to produce the net present value (NPV). The further into the future these occur, the less their NPV will be. Shortening the time also can have a significant impact on the NPV.

The following example illustrates these points for an infrastructure investment of £200 million matched by a vehicle supply cost of £40 million, with a 7% discount rate. In Table 5.6, construction is spread over 10 years, in Table 5.7 over 5 years, and in Table 5.9 over only 2 years. Thus, from Table 5.10:

$$\frac{\text{NPV revenue}}{\text{NPV cost}} = \frac{24.9}{175} = 0.14 = 14\%$$

From Table 5.11:

$$\frac{\text{NPV revenue}}{\text{NPV cost}} = \frac{43.3}{207.2} = 0.20 = 20\%$$

From Table 5.12:

$$\frac{\text{NPV revenue}}{\text{NPV cost}} = \frac{50.8}{231.6} = 0.22 = 22\%$$

In these three examples, the DCF technique shows that completing the project quickly and starting up passenger revenue service early is the most profitable option, but it needs an annual capital expenditure to increase from £20 million to £100 million. For a real project, many more options are likely

Table 5.11 Project DCF with 5-year construction period

	NPV cost		
Year	*Infrastructure*	*Vehicles*	*NPV revenue*
1	40	0	0
2	37.2	9.3	2.0
3	34.6	8.5	3.7
4	32.2	8.0	5.2
5	29.9	7.5	6.4
6	0	0	6.0
7	0	0	5.5
8	0	0	5.2
9	0	0	4.8
10	0	0	4.5
TOTAL	173.9	33.3	43.3

Table 5.12 DCF with 2-year construction period

	NPV cost		
Year	*Infrastructure*	*Vehicles*	*NPV revenue*
1	100	20	2.0
2	93.0	18.6	4.5
3	0	0	5.9
4	0	0	6.0
5	0	0	6.4
6	0	0	6.0
7	0	0	5.5
8	0	0	5.2
9	0	0	4.8
10	0	0	4.5
TOTAL	193.0	38.6	50.8

to be evaluated. If too many options are considered, there is a danger of "paralysis by analysis." These techniques cannot replace the experience and judgment of a good team that will confirm its conclusions with DCF calculations.

5.6.2 Cost-Benefit Analysis

In the public sector, where capital funds that come from taxes are given as a grant, there is a danger that this "free money" might be wasted, so government finance ministries use a technique called cost-benefit analysis (CBA) to evaluate different projects. CBA uses a methodology similar to DCF in the private sector, but includes impacts that have costs or benefits not captured as cash transactions.

Sometimes these impacts can have indirect financial implications. For example, improving safety in transport will mean lower medical costs, perhaps even costs associated with maintaining people who are disabled after a transport crash. There also may be other benefits, like fewer lost working days. None of these produces any cash transaction for the transport project, but will clearly benefit the social services budget of the government or other public agency and improve workplace productivity. Therefore, economists create proxy (money) values for these costs and benefits that are then included in the CBA.

Where the impacts have a cash effect outside the project, then evaluating them is fairly easy. The difficulty arises when impacts do not have a cash transaction (e.g., damaging the habitat of an endangered species that may become extinct). Here there are two approaches. The first is the cost of doing something (different) to avoid the anticipated impact. The second is more subjective and requires market research to find out how much people would be willing to spend to avoid the impact (e.g., by moving to a location where there is less traffic/noise/pollution).

Similarly, some of the cashless benefits might fall on the users of the project (e.g., saving travel time). The users are not asked to pay directly for these benefits, unlike a commercial project, where the users pay for the service they enjoy. How can these benefits be evaluated and included in the CBA?

Two basic approaches have been adopted. In the first, an attitudinal approach is used, where people are asked how much they would be prepared to pay to, for example, save time. Conversely, they can be asked how much longer the journey time can be if the cost of travel is less.

The second is a behavioral approach. It studies people's travel choices to learn what options, when available, they choose either to save time or money. This approach is more difficult but gives more credible results. On the other hand, the first approach is easier to undertake but will need behavioral information to validate.

In either case, the two dimensions of time and money can be plotted on a utility curve and will give a range of "value of time" figures (Figure 5.7). For simplicity, the average is normally used. From considerable research, there are big differences between income groups and journey purposes.

The work journey, because it is economically important, usually is valued, from the kind of research outlined above, at about the wage level of the traveler. This goes some way to explain why the private car dominates the

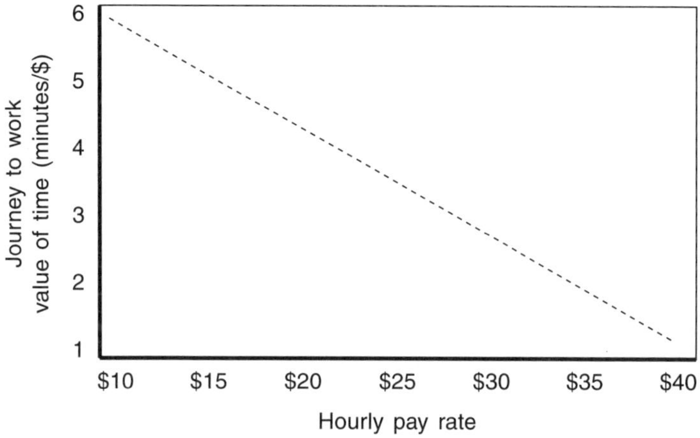

Figure 5.7 Value of time (Graph: L. Lesley)

journey to work market even though car travel is much more costly than any form of public transport, including taxis. Other journeys, including leisure, normally are lumped together and valued at about 25% of the wage rate, although there are differences for different journey purposes.

For most transport projects, the main benefit (about 70%) evaluated for CBA comes from travel time savings. As most transport projects are dominated by local trips, the time savings will be only a few minutes. This raises questions for the CBA. First, can such small time savings be detected, especially when variability in travel time is large because of congestion? Second, if detectable, can the time be put to some economic use? This is one reason why CBA in many European Union countries includes time savings greater than 15 minutes, which can have an economic use and hence value.

When a large number of very small time savings can be used to justify spending tens of millions of public money on a transport project, the CBA becomes very sensitive to the capital cost assumptions. Even at the most confident of capital cost estimates, CBA shows many projects to be barely viable, especially those dependent on travel time savings to justify.

It therefore only takes a small capital cost overrun to make the CBA negative. By this time, the project may be too far advanced to stop. Hence the severe criticism by the UK National Audit Office in 2004 of the publicly promoted and funded light rail projects that failed to generate the number of passengers predicted and their time savings and revenue. Capital cost overruns had often resulted in cutting corners, sometimes leaving a poor quality or incomplete system.

5.7 Implementation and Phasing

To avoid the problems identified above, the project manager must be very confident in the plans and cost estimates for the project. The project manager also must ensure that the phasing is such that time overruns do not turn into cost overruns. The key to a smooth and successful new project is confidence and trust amongst all those involved in the construction, installation, commissioning, and operation of the new light rail line.

The project manager representing the promoters must have confidence that the contractors will be able to do the job in the agreed time, quality required, and within budget. The contractors must be able to trust the project manager that the contract details will not be changed (significantly) during construction and that payments will be honored in the stages agreed. This is important because most of the contractors that fail do so not due to lack of profitability but because of a cash flow crisis.

In the European Union and UK, major projects which are funded by the public sector are required by EU directives to have open and accountable spending. This mandates that all such projects (valued at more than a small cost threshold) are advertised in the *EU Official Journal* and that a system of competitive tendering is employed. One of the strategies used by contractors is to price a project very low and rely on claims for contract variations to generate the profit required.

This was another area where the UK National Audit Office in its 2004 report criticized new light rail projects in Britain. Indeed, the logic for the adoption of design, build, operate, and maintain (DBOM) contracts, also known as "turnkey," is that the risk appears to be transferred to the private sector contractors, since any cost overruns will eat into their profits. In practice, things have not always turned out like that.

Another criticism of the UK National Audit Office was that every new UK light rail project was implemented using a different contractual arrangement with the private sector suppliers (Figure 5.8). As an example, the South Yorkshire Supertramway was a design build with one contractor and operate and maintain with another. When a different contractual arrangement is used to implement each light rail system, there is no opportunity to gain experience or realize improvements. In contrast, most roads are constructed according to a standard contract, so the promoters and contractors have built up experience, and this rarely creates problems in implementation.

Light rail systems in Britain reviewed by the UK National Audit Office all had implementation and interface problems arising from these different contractual arrangements in which no party had previous experience. Needless to say, this generated a lot of work for lawyers, which the UK National Audit Office observed took a larger part of the budget than the engineers who designed and supervised the entire system build. In any case, there is a logical

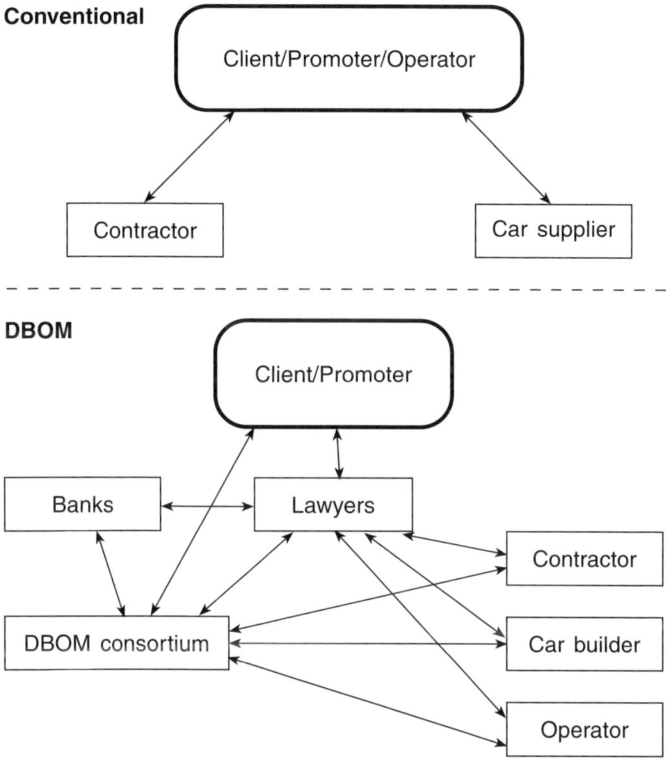

Figure 5.8 Two contracts compared (Drawing: L. Lesley)

anomaly with all these arrangements that revolves around the assumption that the risk is transferred to the private sector.

In all projects promoted by the public sector, the biggest risk is the public sector itself (e.g., change in political control, new regulations, etc.). As the private sector cannot possibly anticipate what these public sector risks might be, they have to be factored in at a higher price.

The engineering risks of constructing a light rail or tram system are well understood, with more than 100 years of experience worldwide upon which to draw. Similarly, the risks of operation on the supply side also are well documented. On the demand side, competition from private cars, taxis, etc. can be modeled in the financial assumptions, but with more difficulty, as they are subject to public policy for fuel duty, parking charges, etc. The private sector factoring in public sector risks is one of the reasons why, on a comparable basis, UK tramway projects have cost twice as much as those in other European Union countries where conventional client/contractor arrangements are used.

Where there is a conventional contract, the public sector accepts the risk from its own decisions (e.g., new specifications, stations, etc.), while the private sector accepts and warrants its contractual responsibilities for workmanship, health and safety of its staff and the general public, and completion on time. The demise of Metronet in London in August 2007 was primarily due to the public sector changing the specification and then not compensating the private sector contractors adequately, which led to their going bankrupt. Both parties lost. Indeed, the fallout from such collapses is that private contractors become wary of doing such work for public bodies.

Another example of public sector risks is the high cost to the private sector of bidding for projects. Typically, a contractor will spend about £2 million to bid on a £300 million light rail project in the UK and perhaps win one in five bids. This means that £10 million is spent on average to win a contract, which of course has to be factored into the bid price. Worse still, some projects are canceled after the contractors have invested so much in submitting bids.

In 2006, public sector tramway projects in Leeds, Liverpool, and Portsmouth were canceled (Figure 5.9). One of the results was that AMEC, a major civil engineering contractor, pulled out of any further public sector work after having spent significant sums on abortive bids. Another casualty was Laing's, which went bankrupt due to having to write off unsuccessful bidding costs.

There is no doubt that a straight contract, where the promoter specifies what is to be done and the contractor builds it, is the least complex arrangement, as well as the one most commonly used throughout world. It is well understood by the promoter and the construction industry. Here, the risk of the promoter changing specifications is assumed by the promoter, while the contractor assumes the business and performance risks of completion on time, to specification, and within budget. As these aspects are all under the control of the contractor, there is no need to build into the bid price any extra cost elements. Construction risks, even variable ground conditions, which can have a big impact on foundation costs, can be factored as a normally expected risk rather than a "super risk."

The big challenge for the contractor installing a new light rail system will be interfacing with a large number of outside bodies, which are neither the client nor system users. This is because rail construction work is spread throughout a city rather than on one constrained and exclusive or private site. Organizations that will have an interest include the chamber of commerce representing merchants along the route, whose access or trade might be affected during construction. Utility operators (gas, electricity, water, telecommunications, wastewater drainage, highway authorities, etc.) all will interact with the contractor. Good communication and understanding before

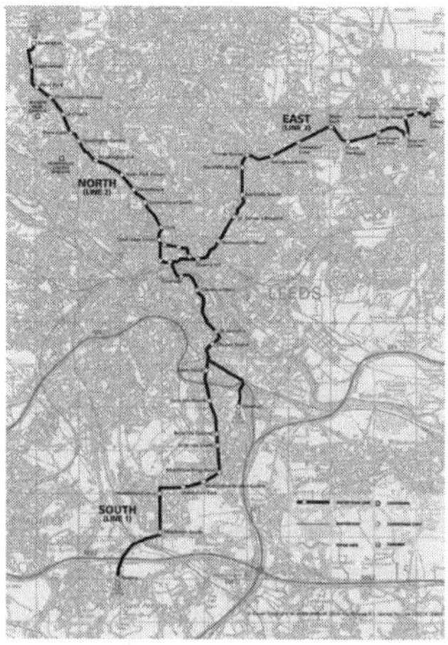

Leeds Supertram

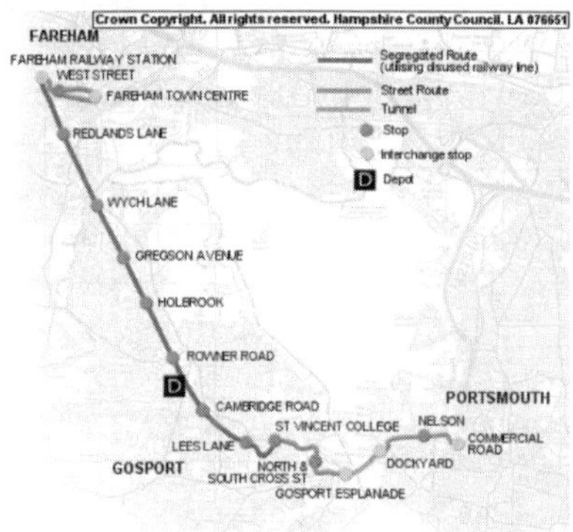

South Hants Rapid Transit

Figure 5.9 UK projects abandoned in 2005 (Photos: Merseyside Passenger Transport Executive, West Yorkshire Passenger Transport Executive, and Hampshire County Council)

going onto the site will be important for the smooth and problem-free implementation of the tramway.

One way to achieve this is a consultation forum where technical issues can be addressed, especially in instances where satisfying one utility might impact others. To make this work, the contractor needs to have clear technical issues to discuss that cannot be resolved on a one-to-one basis with the utility companies; otherwise the forum is likely to degenerate into shop talk, wasting everyone's time and reducing the credibility of the contractor.

Similarly, establishing good relations with the local community, whose neighborhood will be disrupted even if only for a short while, will be key to smooth implementation. For example, it may be necessary to temporarily block access to some houses on a particular street. With sufficient planning, the contractor might offer to pay for the residents to stay in a hotel during the critical time and guarantee to provide security for their homes. This approach might be easier and better for public relations than having to work around and severely disrupt the lives of the residents for some weeks.

Another aspect of implementation is opening the system for passengers. The critical factor here is the location of the depot, since rail vehicles need to be stabled. It may, therefore, make sense to build the tramway nearest the depot first, to enable both staff training and commissioning and, as importantly, to begin passenger service on a small scale so that any operation problems can be resolved before the full system is opened and the system can start to generate revenue, which will be important for the operator's cash flow. This means almost certainly that construction of the tracks through the city center will take longer, as it is a more sensitive environment.

Bringing materials to construction sites is a logistical problem that will have to be resolved. If the light rail is built on a moving front basis, then it might be possible to use rail cars to deliver material to the site from the storage depot. On the other hand, material deliveries can be on a "just in time" basis, where heavy road vehicles deliver directly to the track construction site without the need for intermediate storage and handling. If this can be arranged, it is likely to impact traffic in the city the least (Figure 5.10) and therefore should be the most acceptable approach to residents and businesses along the route.

In any rail installation, there is preparatory work, which builds on surveys undertaken at the design stage, to ensure that no subsequent changes have been made (e.g., to street furniture such as lampposts). These changes might

Figure 5.10 Track construction in Helsinki (Photo: M. Bell)

not impact the rail system, but they could affect the planned method of construction.

The next stage is diverting traffic to provide a safe working environment for contractors' staff.

Building a light rail line is intrinsically no more complex than laying a major pipe for gas or water. Some utilities may be physically in the way and have to be diverted. These should have been identified at the design stage and an agreement reached with the utility as to both the new location of the diverted plant and replacement equipment. It may possible to leave some plant *in situ*, but with enhanced access arrangements for maintenance, such as include inspection chambers or modifying existing pits (e.g., for side entry).

The project contractor can choose to undertake all the construction stages or subcontract some or all to specialists, such as track installers, (traction) pole erectors, etc. Either way, there will still be interface problems within the project, which no matter how carefully a critical path chart is prepared will inevitably create hold-ups. Such delays can be due to late delivery of materials or unexpected problems when the road is opened.

When streets in central Manchester were excavated in 1989 for the Metrolink, large cavities under the road were discovered where sewers had collapsed and washed away the subsoil. The only thing holding up the road surface was the tram lines abandoned in 1949 and left *in situ*, which provided short bridges. Fortunately, ground-penetrating radar can now locate such voids before excavation.

Finishing the construction and getting ready for commissioning will invariably produce a list of things not done correctly that need to be rectified. The contractors will, of course, try to ensure that everything possible is done right the first time. This saves time and considerable cost, in some cases enough to mean the difference between profit and loss. Where a problem lies with a subcontractor's work, there may be interface issues like cables not run through the right duct.

The final check with the client will be a walk through the system, possibly also with the licensing or approval authority, which may be required before any vehicle operation can begin. This approval process varies from country to country, and in some countries the operator self-certifies compliance of the system with health and safety regulations. In this scenario, if something goes wrong, the operator will be held fully accountable unless third-party negligence can be proved.

5.8 Revenue Operation

Before revenue operations can begin, the system, or the part that is ready, and its operating and emergency procedures will need to be demonstrated, in many countries to a licensing authority. As part of this process, emergency services will be briefed and, where appropriate, training sessions provided in, for example, rescuing passengers from a crashed light rail vehicle. The traffic police and highway authority will have determined the traffic management and control systems, including light rail vehicle operations and, where installed, traffic preemptions and road priority. This will ensure that no dangerous conditions are created and that the traffic plan works as expected.

This control exercise also will include emergency plans should, for example, the power fail and traffic lights not work and light rail vehicles stop running, even if only for a short period. Emergency plans also will be drawn up for access by ambulances and rescue vehicles to get to the scene of an incident and then evacuate any casualties to a hospital. Similarly, there will be emergency practices for problems not on the light rail line, but in and around the area (e.g., an adjacent building catching fire).

While this may seem like a lot of work, research (Cuffe 1989) has shown that transport system controllers cannot improvise during emergencies without making the situation worse. Drawing up and practicing a variety of emergency plans is, therefore, a key part of the preparation for revenue operations. Fortunately, most of the functions on a light rail vehicle are fail-safe, so that if there is a medical emergency with the driver, the vehicle will stop. Of course this might block a major junction, but another staff member will be able to move the vehicle to a safer location so that the driver can be transported for medical treatment. The probability, however, of this kind of emergency is extremely low, especially with appropriate driver training and regular medical examinations.

After emergency routines have been developed and practiced, the next stage is to practice the normal service, but without passengers, usually for between a week and month, depending on whether it is a new system or a new line on an existing system. This practice operation without passengers allows staff to become fully familiar with the service, stopping distances, traffic junctions, and maintaining a reliable timetable.

During this period, the operator might have groups of residents (e.g., elderly, disabled) try the system out. Most new systems have at least one day, often during a weekend, when residents are invited to ride the service for free. With less other traffic and light rail vehicle passenger flows high, this is a good and final "dress rehearsal" prior to the start of revenue service the next day. The goal of the service should be to attract the maximum number of people from their cars, to provide safe and reliable service, and to help passengers gain confidence in traveling by light rail vehicle.

For the operator, the only other important aspect is ensuring that passengers pay for their travel. Clearly, the fare system used will have a big impact on how this is achieved (Figure 5.11). Historically, many urban passengers have paid according to length of journey by graduated fares. In commercial

Figure 5.11 Example of innovative ticketing (Photo: Calgary Transit Authority)

terms, this has a large drawback in that it does not fully reflect the cost of providing the service. The marginal cost of running at cruising speed is very low, whereas the cost of stopping and starting represents at least 70% of the operating costs.

American transit operators realized this in the 1920s when converting from a conductor collecting fares to a one-person operation. At the same time, a single-price ticket (flat fare) system was introduced. This more closely reflected operating costs and also made it easier to ensure compliance, since a passenger either had a ticket or did not, with no grey area.

Similarly, labor shortages in the late 1950s, which made it difficult to recruit tram conductors in West Germany, led to the introduction of a one-man tram operation, again with flat fares introduced. Compliance was encouraged through a penalty fare, typically 10 times the flat fare, and then inspecting 1 in 10 tickets. The last urban system in West Germany to change was in the newly created Rhine Ruhr Transport Cooperative area in 1978. Graduated fares were replaced by a system of flat and zonal areas that covered the entire system in the Ruhr (www.vrr.de/de). At the same time, the average fare was increased by 10%, which on the basis of previous experience should have reduced patronage by 3%. Instead, patronage increased by 4%, and hence revenue increased by more than 14%.

Choosing the right fare system is an important part of market research for a new tramway. Selling travel by time is another way in which many European systems operate. A ticket for unlimited travel for one hour, four hours, a day, several days, or a week is a simple way to ensure compliance. For the passenger, the price of extra travel is zero during the time period for which the ticket is good, and the operator's cost to carry an extra passenger is effectively zero when there is spare capacity in off-peak periods.

Low-cost airlines have revolutionized air travel with cashless transactions and ticketless journeys. In Europe and North America, virtually everyone now has a cell/mobile phone. In the UK, there were 1.3 mobile phones for every person in 2007. Many services are available through mobile phones, and there is every incentive for new light rail operators to use them to collect fares.

Once a phone number is registered, a passenger profile can be built to give information and to sell additional services, sometimes in association with activities and attractions in the area. Very few people ride light rail vehicles just for pleasure; most are going somewhere to do something. Linking travel with store promotions or discounts for movie, sporting event, or theater tickets can be a way to generate extra patronage and revenue, especially during the off-peak when there is spare capacity and marginal costs are nearly zero.

Similarly, passengers willing to make a travel commitment some time in advance get a better price than those who show up at the last minute, as low-cost airlines do. Indeed, it is likely that a simple single-journey ticket machine, located on every platform stop, will offer a cash ticket at about five times the price of one bought with a cell/mobile phone.

Monitoring patronage is a critical part of maximizing revenue. This should be done on a daily basis, and the traffic manager must know first thing each morning what traffic was carried the day before. If there is any significant deviation from sales targets, programs to encourage usage can be implemented, such as targeting passengers (by mobile phone) who live in a particular part of the system by offering special fares and issuing news releases to the local media to raise awareness of the service. Promotional offers also can be used, especially in the off-peak when there is spare capacity. This is an indirect form of passenger/customer feedback.

The operator also should solicit direct and regular feedback from customers and, as importantly, potential customers. Obviously, the operator needs to retain existing passengers as the first priority. Encouraging them to become more frequent riders increases revenue. As important is finding out why potential passengers do not ride. The reason may be related to perception of the quality of the service or not knowing the fare system or timetable, both of which can be addressed by a public information campaign, especially a directly targeted one.

6

Marketing and Advertising

Marketing is often confused with advertising. The two are quite separate activities, and advertising should follow marketing, not the other way around. The purpose of marketing is to find out what (potential) customers want. The purpose of advertising is to tell (potential) customers what is being offered. Both, however, are aimed at increasing sales and revenue.

6.1 Marketing

Given that a commitment to an investment in a light rail system has been made, clearly a marketing exercise will not be aimed at finding out if residents would prefer to use their cars or buses. Rather, at an early stage, marketing should seek to determine the best routes in terms of potential riders and traffic attractions (Figure 6.1). It also can establish or confirm the travel patterns of residents and therefore the trips made that could be attracted to the new tramway. Similarly, marketing can focus on such details as color scheme, brand name, the fare system, and customer care.

A light rail system can mimic car manufacturers, transit's principal competitor. The launch of a new car is preceded by painstaking feedback from target customer profiles. This includes workshops on details like the shape, lights, fittings, and other features that customers use to differentiate one car from a competitor. Similarly, potential light rail passengers can be asked, in a systematic way, their preferences for different aspects of the vehicles, the stops, and service. Use of indirect methods to assess attitude is likely to be

- Reach and frequency analysis
- Socio-demographics of passengers
- Frequency of trip making
- Incidence of overseas Filipino workers in families
- Type of light rail transit ticket used
- Time of journey making
- Time spent riding light rail transit
- Products most often used
- Exposure of other media
- Recognition of advertisements in stations
- Number of passengers per station

Figure 6.1 Research topics covered in Singapore light rail transit travel habits survey (Source: L. Lesley, based on project undertaken by Neilsen Media in May 2009)

more accurate than direct methods, where the responder may try to give the answer he or she thinks the questioner is seeking.

Like carmakers, the light rail promoter also should use mock-ups to assess the ergonomic ease with which passengers can use the new system and the attractiveness of different aspects, including coverings and colors. Similarly, getting feedback on navigation through the system, including information needed that should be provided, can be a salutary lesson for the promoter since professionals make assumptions about people's ability to understand for the first time a new travel process in a strange environment.

In order to make this exercise realistic, some design work will be needed, for which a cost is involved. Getting it right in public relations and market acceptability terms is very important for the financial success of the system. Marketing costs are small compared to the major investment in building the system. Ensuring that responses to such exercises are valid in their own right is also important.

The science of drafting customer feedback exercises is well established, although of course the product being tested each time normally is different, and too few light rail systems are customer tested before being built. A preliminary stage in the form of a pilot exercise is a simple way to ensure that the main study will be valid. The pilot exercise allows wording and understanding to be checked, as well as the emphasis of the survey to be rebalanced, from the feedback. Spending in this early phase can avoid wasting ten times the money or more later and is much less than the cost to retrofit or change equipment, signs, etc. after the system has been opened.

6.2 Advertising and Launch

The experience already outlined of the launch of the BEELINE BUZZ Company in Manchester and the reaction of existing operators show that, in the words of Prof. Marshall McLuhan, "the medium is the message" (McLuhan 1967). Often, low-cost direct advertising is more effective in raising (potential) customer awareness than expensive television commercials. This is particularly so for travel decisions, where more than 70% of all journeys start or end at home. Having access to valid and easy-to-use information at home will build confidence in the system. Access to information away from home is now available through cell/mobile phones. An SMS text messaging service should give the normal operation times and, as importantly, any out-of-the-ordinary information like service disrupted by traffic accidents. This will keep passengers informed and reduce their stress and frustration.

In 2001, Nokia ran a pan-European television commercial that showed children being alerted by their cell/mobile phones while they were playing at home in Helsinki. They put on their winter clothes and rush out just as a tram approaches. They throw snowballs at it. In Britain, the commercial was withdrawn after pressure from bus operators, who feared it might incite children to throw stones at buses.

This, of course, missed a number of points. The most obvious was that no British bus operator, at that time, had a (real-time) text messaging service advising the approach of a bus. Perhaps the bus operators also were concerned that people might associate the "upscale" Nokia not with their buses but with trams, which were not available in most of Britain and no bus operator was thinking of converting to them. As for other consumer products, targeting the advertising to the appropriate audience by segmenting the market will both be more cost effective and achieve better results.

Children influence parental decisions, but getting the message across to children can be difficult. The building of a new light rail system can be preceded by a campaign of talks at schools about the "dangers" of the tracks and at the same time positive messages about green, safe, convenient, and comfortable travel. McDonald's did not get as big as it is by targeting senior citizens. Tokens or free tickets for families can be distributed through schools, usually for a ride on the Sunday before the start of revenue service. Similarly, groups of schoolchildren can be invited to visit the depot and see how the system runs.

Teenagers are the most mobile section of the population; on average, each makes about 25 journeys per week, nearly 20% above the average. Teenagers also have disposable income. While teenagers often get bad press for rare instances of bad behavior, most are well mannered and active members of their communities. Recruiting teenagers as committed rail riders (Figure 6.2),

> **Children 'prefer buses and trams'**
>
> Four out of 10 teenagers are ditching a lift in a car for the bus or tram, according to a survey.
>
> *20th June 2006*

Figure 6.2 Teenage travel survey in London (Source: http://news.bbc.co.uk/1/hi/england/london/5098872.stm)

before they get their driver's license, is important in building a good and permanent market share.

Incentives like frequent rider rewards (e.g., T-shirts, tickets to pop concerts, etc.) are an economic way to build loyalty. Today, the most effective way to communicate with teenagers is by cell/mobile phone. Teaser campaigns, as used by the BEELINE BUZZ Company, raise awareness and build the market, as does giving promotional T-shirts to teenagers.

In most European countries, senior citizens travel free or at reduced fares. In the UK, operators are rewarded not on the number of seniors carried but on a "no-better/no-worse" formula, the basis of which is that the operator only receives the same revenue generated by (fewer) seniors paying the full fare. Many senior citizens, however, have an important family role of baby sitting and watching children.

A light rail operator that accepts senior citizens' concession travel can develop patronage by offering reduced fares during the off-peak for children traveling with them on the basis that because there is spare capacity, the marginal cost of carrying extra passengers is low. Generating extra revenue from accompanied children is worthwhile since it brings in more money than a near-empty vehicle would. Children who ride with their grandparents are likely to persuade their parents to ride the system.

Senior citizens are the least mobile section of the population. Because they do not go to work, they take 10 fewer trips than the average 21 trips made per week. Because senior citizens over 80 are often effectively housebound, enabling them to ride more often not only makes good commercial sense but also performs an important social function. Many senior citizens and certainly those close to retirement are car owners and drivers. The "treat" of a rail ride with grandchildren can be a way to help reduce their dependency on car travel.

Generating such accompanied trips requires the operator to liaise with cinemas, theaters, museums, and other places where events for children are organized and offer package deals for children and grandparents or other adults to ride the system and attend such events. The use of the Internet and cell/mobile phone ticketing makes this easy to administer at a low cost but with high-margin revenue.

Such packages should be promoted by both the rail operator and the venue. The venue can even offer "free" travel with the event ticket. Here, the whole is greater than the sum of the parts, as both participants benefit. Entertaining children away from their home also has social benefits. Parties for schoolchildren can be organized in the same way, since a class of children can easily be transported in a 200-passenger-capacity light rail vehicle without overloading it. For schools, not having to charter buses is economic and offers flexibility.

Numerically, the largest travel market is adults between 21 and 60, who are also the most committed car users. Here, advertising directed at adults has to address the reasons why people drive instead of using transit:

- Convenience: A light rail system with high service frequency offers a show-up-and-go service, without the chore of having to find a parking space (and paying for it).
- Speed: In urban areas, a light rail system with priorities offers a faster end-to-end travel time, especially during peak hours, than driving a car. Passengers can use their cell/mobile phones legally on trams and read the newspaper, which is not possible in a car.
- Reliability: Electric light rail vehicles are the most reliable road vehicles. They rarely run out of fuel, crash, have dead batteries, etc.

Only by riding the system will people build up confidence in its level of reliability. Word-of-mouth recommendation is the most powerful advertising. This is an important way to spread the message. Indeed, personal recommendation is often the most cost-effective method of advertising.

Some companies employ sales staff to frequent places used by target customers, like pubs and clubs, to spread the message by word of mouth. This also is a good way of getting feedback, at least on perceptions, and therefore an opportunity to dispel myths. This is a form of active market research. Similarly, complaints can be turned into a marketing tool. The data on the way complaints are handled shows that a positive response from a company has a positive and longer lasting impact on the person who made the complaint and his or her friends.

Cost is often cited as a reason for not using public transport. In practice, as many surveys have shown, driving a car is more expensive, although most people do not know exactly how much it costs. Typically, 14% of household income is spent on owning and driving cars, but the perceived marginal cost is low. Because the gas tank is filled infrequently, the cost is not equated with the daily journeys made.

Most high-ridership European tramway systems sell their services in units of time (an hour, day, week, month, and sometimes a year). If coupled with direct bank payment and the mailing of each new ticket, along with promotional material, light rail riding will have an even lower perceived marginal cost than driving. Also, all extra rail journeys have an effective zero cost to the passenger.

An interesting European experience in promoting tickets aimed at car commuters is the so-called "Rainbow Ticket." Basel, Switzerland opted for an economic fare and budgeted on a slight loss from the extra capacity needed. In practice, nearly 100% more drivers than predicted signed up for the "Rainbow Ticket," and the operator made an operating surplus, as well as significantly increasing total patronage, reducing peak-period motor traffic, and reducing the journey time of light rail services.

6.3 Building Patronage

In a pessimistic business plan, the operator will assume no fare-paying passengers on the first day of revenue service. In an optimistic plan, 50% of target patronage can be expected. Whichever is the starting point, the operator will work on the basis that patronage will increase as advertising works, word-of-mouth experience is shared, and promotional efforts are rewarded. Obviously, no light rail system can expect to achieve a 100% market share, but 20% of the trips made in the light rail corridor is a modest target. There are examples of European cities that get nearly 50% of such corridor trips.

If a tramway replaces or supplements an existing bus service, what patronage can be expected compared to the original bus service? There is considerable evidence on which to base this assessment. Probably the most systematic was US Transportation Research Record 1221 (Tennyson 1989). Two different approaches were to taken to assess the patronage that a light rail service might expect compared to an equivalent level of service with buses. The first set of data came from the replacement of rail and streetcar lines by bus service (Figure 6.3). The second came from the replacement of bus service by light rail. Both show that a light rail system can expect to carry between 35 and 44% more passengers than an equivalent bus service, in terms of frequency, stopping places, speed, fares, etc.

More recently, a number of new light rail systems have opened in France. They usually replaced the busiest bus routes, and the bus network was reorganized to feed into suburban rail termini. Here again, the new tramways carry at least 30% more passengers than the bus routes replaced. Interestingly,

Figure 6.3 Rail to trolleybus in London, 1938 (Photo: London Transport)

the bus network in the same towns also experienced an increase in patronage of a similar order, apparently sharing in the perceived improvement in the quality of the network and, of course, from extra feeder trips.

Finally, in the UK between 1950 and 1960, virtually the entire intense network of urban tramways was abandoned. In some places, the tramways were replaced by trolleybuses, in most diesel buses. In all cases, the replacement buses were new, often faster, and ran more frequently than the "inflexible" trams they replaced. The result was almost an immediate drop in patronage of the order of 30%. More research was needed to determine why, but was not undertaken at that time. The public sector operators were complacent about their service and the fact that passengers did not own cars and therefore were dependent—but not for long, as the abandonment of rail service became a catalyst for buying cars. The same experience was true in North America 10 or more years earlier.

Market research on public services was almost unheard of in those days, and had this not been the case, many tramways would have survived in Britain and France, as they did in Germany, to be modernized later. Why? There was a lot of public support for tramways, and indeed, in many UK cities, petitions were drawn up against the abandonment of tramways. In

1948 in Liverpool, a petition attracted 250,000 signatures, more than half of all voters, but was ignored by the city council and pushed through on the vote cast by the mayor.

On the other hand, a new light rail system should not be built just to divert bus patronage. The target market must be private car trips for two reasons. First, they are the majority in the cities of all developed countries. Second, reducing car traffic will create the space needed to allow light rail systems to have priority, in terms of both road space and junction time. This will ensure that new light rail systems remain an attractive alternative. In community terms, reducing air pollution, greenhouse gas emissions, noise, and the need to import oil cannot be ignored. These factors can be an important part of the justification for new publicly funded systems.

Building new patronage initially should concentrate on the people living and working in the light rail catchment area. In many places, this will be those who can "walk and ride." If park-and-ride stations are also provided, as they should be on all lines, then suitable road signage for drivers coming into the city, especially if linked to free-flowing lanes into park-and-ride stations where the main road is congested, will attract out-of-town commuters.

Advance advertising is certainly a strong part of the McDonald's launch of all new outlets. Indeed, McDonald's shamelessly targets children on the basis that their parents will be persuaded to come along, again and again. Capturing customers for light rail at a young age also should be part of a new public transport service. An outing by rail will be a new experience for most families who otherwise go everywhere by car.

An upscale image as well as high-quality service need to be presented. Families must be attracted by something better than their car, at least for traveling around town. Influencing local residents requires a combination of direct information at home and incentives at the places most often frequented, including work, shops, entertainment, etc. Indeed, a free first day, often a Sunday, can be used to get families who normally only travel by car together to try a rail ride. Special attractions in the city center could be the incentive to ride the rail, if a joy ride in its own right is not enough. The novelty of a new light rail system should not be underestimated.

6.4 Product Life Cycle and Relaunch

All consumer products are periodically relaunched or die. A new light rail or tramway system is no exception. The questions are when the best time for a relaunch is and how it should be done.

One example is the South Yorkshire Supertramway, which opened in 1995. In 1998, it was sold as a loss-making concern to the Stagecoach Group on a 20-year concession. In 2000, Stagecoach made operational changes that

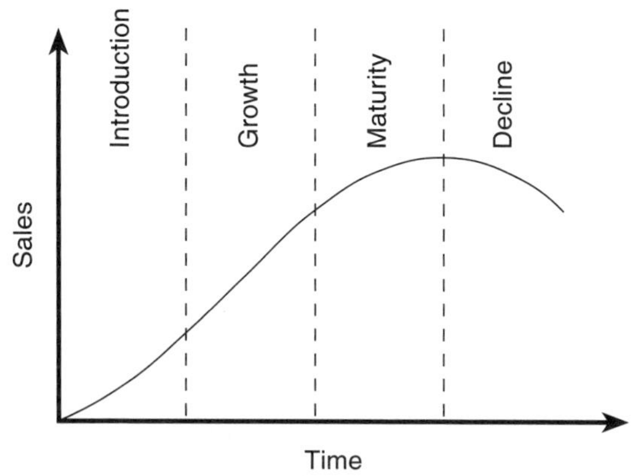

Figure 6.4 Product life cycle (Graph: L. Lesley)

included replacing prepurchase station ticket machines with on-board conductors to collect fares.

Stagecoach then undertook a rebranding and relaunch (Figure 6.4). Initially, this concentrated on a new livery and logo based on the Stagecoach house style. The original livery and logo were an overall grey; the new Stagecoach livery and logo are white with orange and blue stripes. This exercise in 2001 was undertaken after operational changes had produced an increase in ridership. The relaunch reinforced this growth, resulting in a continuing patronage increase that took the annual traffic from 8 million in 1998 to 12 million in 2006 and an operating loss to profitability.

Subsequently, the vehicle interiors were refurbished and another livery and logo applied from 2007. This signals a new system creating its own positive stamp on the transport market of Sheffield. The 25 light rail vehicles on the 26-km Supertramway now carry 10% of all transit trips in the region, as bus use has continued to decline.

This example provides some clues as to when and how to successfully relaunch a light rail system. After opening, patronage will grow, but will reach a plateau after a few years. If nothing is done at that point, patronage will begin to decline slowly. Judging the optimum point at which patronage plateaus is the key to a successful relaunch. Market research is used to find out the good things about the service and the things that deter nonusers. Addressing those issues is the key to building new market growth.

Stagecoach had strong feedback that passengers did not feel secure in one-man-operated tramcars. The introduction of conductors immediately increased the feeling of security. It also produced a one-off increase in revenue of the order of 7%, as nearly all fares were collected. Indeed, this increase in revenue coupled with the savings from not having to repair vandalized ticket machines made the conductors self-funding.

The timescale for introducing a systemwide relaunch raises the issue of tactics. On a small system with few vehicles, a complete change can be planned and implemented almost at once. On a more complex multiroute network, a route-by-route relaunch will be the most effective approach, starting with the busiest or most profitable route. Then, the extra profits from rising revenue can be used to fund the subsequent route relaunches. Such relaunches can coincide with the introduction of new vehicles, again usually on the busiest route, with a cascading of older vehicles throughout the network and the retirement of the oldest from the quietest route. This at least gives the impression that all routes are getting "new" vehicles.

FirstGroup has used the brand name "Overground" (Figure 6.5) in relaunching some of its bus networks in large UK cities like Glasgow. This cleverly plays on the reputation of the (London) Underground for speed, frequency, and reliability of service. No information has been published on the effectiveness of this in relaunching bus service since the route networks also were changed, from fine networks with infrequent service to coarser networks with more frequent service, in order to achieve shorter waiting times.

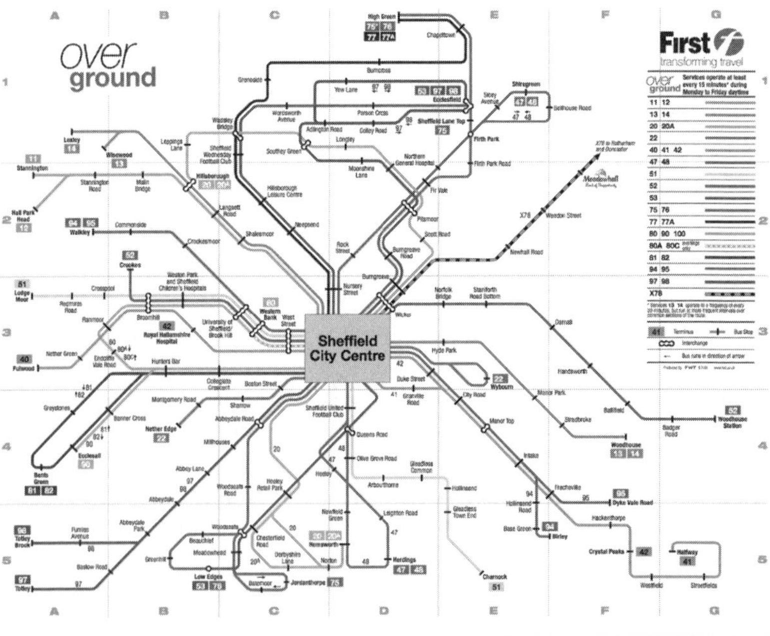

Figure 6.5 FirstGroup Overground in Sheffield (Photo: FirstGroup PLC)

Products can be relaunched for other reasons. A classic was the formation in Germany of the Rhein-Ruhr VV (VRR) in 1978. This is a voluntary association of many public transport operators in the area. The Rhein-Ruhr area, which is as big as Liverpool and Manchester or St. Paul and Minneapolis, with more people, also had almost 100 public transport operators. The staff needed to coordinate these 100 operators was only 24.

The logic of the VRR was to coordinate service information and make traveling by public transport easier. The first major reorganization in 1979 was a change from graduated, distance-based fares to a coarser zonal scheme that had the effect of increasing fares on average by 10%. On the basis of previous experience, the budget was based on an elasticity of –0.3, which would give a revenue increase of about 7%. This change was accompanied by a lot of advertising, including posters at every stop. The result surprised the VRR. Patronage increased by 4%, giving an actual increase in revenue of 14%. The conclusion was that the new fare system was easier to understand and, coupled with the intensive advertising, encouraged extra ridership, some from nontransit-using residents. More importantly, light rail operators should not underestimate the power of an advertising campaign for a good product, but should never waste money advertising a bad product.

6.5 Staffing

Public transport (transit) is a labor-intensive operation. Typically, more than 50% of light rail operating costs are for wages and salaries. Given the relatively high cost of staffing, it makes sense to use staff as productively as possible and to improve public relations and customer satisfaction. Too often, public transport is run as a production operation, where running the light rail vehicles is seen as the prime objective.

New light rail services are seeking to build patronage from car commuters. Service quality will be an important factor. As the Supertramway in Sheffield discovered, introducing conductors can be cost neutral when the travel environment is improved and new traffic is attracted by the friendlier service and safer riding. Even the fully automated London Docklands Light Railway has "train captains" on every train. They not only collect fares but also give information to tourists and maintain a good atmosphere with security.

When the bus industry in the UK was deregulated and privatized, the established operators used mostly double-deck one-man-operated buses, where the drivers were paid (in 1986) about £3.50 per hour. New operators saw the potential of minibuses to reduce costs.

Instead of recruiting existing bus drivers, these new companies recruited sales assistants from main street retailers who were being paid £1.80 per hour. By offering driver training and a new wage of £2.50 per hour, the new minibus operators were able to recruit staff already good at handling customers. These new operators did not want bus drivers per se, but operating staff trained in customer care. The technical skills of driving a bus are easily learned, especially as most people can already drive a car.

For a public transport operation like a light rail system, the staff that the passengers meet do not just represent safe operations, but are the salespeople, PR office, information service, etc. These extra and important skills are not always fully included in training programs. The other frontline staff are the "inspectors," "revenue protection officers," or some similar title. How these staff conduct themselves can also make or break the image of the service. The city of Brussels found that the local people had nicknamed their inspectors the "gestapo" because of the military-style uniform and their authoritarian modus operandi. This coincided with a decline in patronage—cause or effect?

After a rethink that included new "softer" uniforms and retraining, every inspector appeared as the subject of a poster campaign, with such captions as "Uncle John," "Granddad Peter," and "Aunt Mary" to emphasize the passenger security and assistance service offered. Within 6 months, the public perception of the inspectors and the system had improved and patronage began to grow again. This is another example of an accidental but successful relaunch of a public transport service.

An example of a relaunch that was not successful resulted from the creation of the National Bus Company (NBC) in England. In 1968, all the state-owned and nationalized regional private bus operators were absorbed into a single state-owned company. Many of these bus companies were originally founded by entrepreneurs and were family owned, reflecting the local communities in which they operated.

NBC decided that a corporate image was needed. After the usual advice from consultants, it adopted an arrow logo based on an italic "N" and its reflection in red and blue (Figure 6.6), which gave a symbolic impression of movement. NBC also decided that all buses should be either green or red, since many companies already had (different shades of) green or red. The destruction of many unique and local brands was an expensive mistake for NBC. By comparison, the Greyhound brand name and logo grew out of a nationwide bus system from almost its beginning.

The NBC policy destroyed the loyal identity of most bus companies, which led this author to write an article (Lesley 1974a) in the trade press

Figure 6.6 NBC and Greyhound logos compared

pointing out that many town centers had the same variety of multiple shops and it was becoming difficult to tell urban centers apart. One of the few symbols of civic identity came from the livery and logo of the local bus company. With NBC's new standardized corporate style, that disappeared, and the downward trend in patronage of about 1.5% per year begun in 1956 accelerated to about 2% a year.

6.6 Community Involvement

Today in most towns, very few residents would be able to give a visitor directions to reach a particular place by transit because few people use bus or other public transport services. In Greater Manchester, more than a third of bus trips are made by senior citizens who travel free. Adults who work mostly drive. Even in London, the most public-transport-oriented city in Britain, about 65% of trips in outer London are made by car, and only in the central area does public transport carry more than half of journeys made (Table 6.1).

The involvement of the community as transit service users has declined and with it a second aspect of involvement—civic identity. Many people do not know who runs the bus or transit service. By contrast, most people identify with their local major league team even if they rarely or never attend games. Supporting "their" team is now one of the few remaining local identifiable symbols.

Garnering local support for public transport is going to be a challenge for a new light rail promoter. Without it, priorities, in terms of either exclusive use of road space or preemption of traffic signals, will be difficult to achieve. A new light rail system, therefore, needs a well-planned campaign to generate local support, both so that it will be considered a good idea and in getting people on board as potential customers.

When residents (and electors) see light rail as a facility that will improve their lives and provide better travel options, it is more likely to win community support. Some of the large supermarket chains used to be very good at this. Today, they often face hostility from groups of residents when seeking

Table 6.1 Modal split in parts of London (2001)

London	*Transit (5%)*	*Car*	*Walk/cycle*	*Trips per day*
Central	20	5	75	0.8 million
Inner	20	30	50	4.1 million
Outer	15	55	30	8.6 million

to open new stores. Sometimes the reason is fear that local and familiar stores and district shopping centers will be hurt. When a school playing field is sold by the education authority, to be replaced by a light rail facility, residents fear that children will have nowhere to play.

If people cannot walk to shops, they need either a car or a new light rail system. A new light rail system can form alliances with shopping centers to offer direct and fast rides from the main centers of population in their catchments. Some of the customer car parking areas can be converted into more shops, other commercial ventures, or housing. This is a win-win situation since the light rail operator can guarantee to capture some shopping trips presently made by car.

The above are some of the themes that a new light rail promoter can consider in a campaign to get community involvement, which might also include gaining proponents as investors and ultimately riders.

6.7 Route Structure and Service Frequency

There is an art and a science to designing a light rail system or even a bus network. The starting point must always be the objectives that the network is trying to achieve. Where do people live? Where do they make their regular trips? What other locations can attract new trips? For the passenger, easy access, short waiting times, and fast journeys are important if trips are to be diverted from private cars, which are instantly available. Service attributes are the most important aspect of a (new) light rail system. The fare is usually a fourth or fifth consideration after the service quality. The new operator should be able to offer a fare structure and methods of payment that are acceptable to the target customers. These are demand-side objectives.

Supply-side objectives include maximum vehicle and staff productivity. These are enhanced by a high operating speed, which will also give passengers fast journeys. Normally, vehicle productivity is maximized by having long routes, so that the terminal dwell or layover time is minimized as part of the round-trip time. In a typical city, this means routes that run right across the built-up area through the city center or around the inner core, so that the minimum number of vehicles and crew is needed for the peak service frequency.

On a single light rail line, the journey options for a passenger are "few to few." To maximize the benefits of a network, there also must be interchange stations. These will be between light rail lines and other transport systems (e.g., park-and-ride, taxi-and-ride, bike-and-ride, bus and railway stations, and other new light rail lines when built). Interchanges offer "few-to-many" travel options, but need good travel information and promotion to make realistic.

Figure 6.7 Pico Interchange Station in Los Angeles (Photo: Los Angeles Transit)

The more lines that interchange, the more journey options that are available until a "many-to-many" service network is offered (Figure 6.7). This will benefit, and attract, the maximum number of residents in the area. The location and design of interchange stations are therefore critical. An urban light rail network normally will have high-frequency service of the order of one car every 5 minutes. In this case, timed interchange is not critical. The critical factor is then a short walking distance to reach the new service. A "cross-platform" connection is the best and most convenient.

Second, clear signage is vital so that interchanging passengers know where to wait for the next service. This is particularly important when the very first journey is made, but also is necessary as reinforcement for successive journeys to ensure that the connecting vehicle is going in the right direction.

Interchange is also made easier if through tickets, or any of the family of time-based tickets, are available, because a second payment is then not needed. For the operator, this does not represent lost revenue since a passenger making a journey that can be completed on one line only has to pay once and a journey using two lines has the same utility to the passenger as that on a single line.

Where a second fare is charged, passengers may opt to stay on the first line and walk from the nearest stop or not make the journey by rail at all, but instead drive. Provided the network has some spare capacity, then interchanging journeys impose no extra costs on the operator.

When generated (interchanging) journeys begin to tax the existing capacity, the operator has a number of strategies to reduce overcrowding, short of adding extra capacity. First, alternative interchange stations can be encouraged to spread the demand over more lines. Second, fares can be higher at peak times or lower at off-peak. This may encourage marginal peak-period passengers to travel during the off-peak.

Increasing peak fares was the strategy adopted by the Manchester Metrolink in 2005, after being prevented from increasing the passenger capacity of vehicles by removing some seats in order to let more passengers stand. The third alternative where such overcrowding is regular is to reschedule service. More trams can run in the peak direction only, including skip-stop service, with some then returning to the other terminal nonstop to get back in the peak direction faster.

A new light rail line will increase its utility geometrically as new lines are added, but only where convenient, purpose-designed interchange stations are provided. The tramway operator seeking to maximize market share needs to offer a "many-to-many" service that mimics the convenience of car travel, but offers improved travel speed and avoids the problems and cost associated with parking.

The number of routes/lines will depend on the size of the urban area. The corridors served will reflect the density and distribution of population and existing traffic flows that can be attracted to the new service. This "demand-side" analysis has to be balanced with a "supply-side" analysis of the cost of installing a tramway. There is no point in considering a new light rail route where there might be big capital costs, as for tunnels or bridges. Such high costs would be out of balance with the likely benefits (and revenue).

This is where the art of route planning is important. Experience and precedents from other light rail systems should be tapped to find low capital cost solutions so that a new line is threaded through the urban fabric with a minimum of environmental impact during construction and maximum benefits when operating.

7

Case Studies

A few examples may help to show that light rail can be affordable and lay the foundation for changing the environment, modal split, and travel patterns in a city.

7.1 San Diego, California

San Diego holds the record for the fastest and most economic delivery of a new light rail line in the late 20th century. The first line to the Mexican border (Figure 7.1) was formally proposed in 1978, and 36 months later it opened on time and on budget, including all the design, approvals, and negotiations needed. A good basis had been available in a series of transportation studies that examined different options, including more freeways.

The first (Blue) line shares part of the right-of-way of the Southern Pacific Railroad and took over the San Diego and Arizona Eastern Railroad, which continues to operate a rail freight service to Imperial Valley on a time-share basis with the light rail system. Buying an existing rail right-of-way was a low-cost way of creating a light rail route, but this only worked because it follows a natural movement corridor and therefore guarantees traffic.

Second, the light rail vehicles chosen were "off the shelf." These U2 cars had been developed by Düwag in Germany for Frankfurt am Main, where a tramway was being upgraded to light rail standards. These were well-proven vehicles at the time and resulted in a follow-on contract for the new Calgary system, which chose the same U2 vehicles. They were among the lowest cost vehicles available and allowed Siemens, which took over Düwag, to open an assembly plant in California to satisfy the North American market.

Figure 7.1 Blue line U2 car at the US/Mexican border (Photo: San Diego Transit)

The downside to this choice was that the vehicles had a high-floor design, so that platforms about 3 feet (900 mm) above the rails had to be built at every station. This posed a problem on the railroad right-of-way, where there was a danger of the light rail platforms fouling the wider rail freight cars that used the line. It also required some fancy station designs in order for the downtown part of the line to fit in with the streetscape.

The third cost-saving choice was to have the system entirely on the surface, at grade, thus avoiding the need for tunneling or viaducts. This was made possible by the use of traffic-signal-controlled highway crossings. Later, as demand increased, single-track railroad sections of the line were doubled to allow more frequent service. This later work was undertaken on an incremental basis. The total cost of this 16-mile (26-km) first line was $120 million or about $7.5 million per mile ($4.7 million per kilometer). In 1998, it was carrying 31,000 riders a day, making it one of the most cost-effective light rail systems in North America (Chase 1998). The San Diego network is now 51 miles (82 km) long and carries about 118,000 riders per day or about 37 million annually, a density of 0.73 million riders per mile of network.

The San Diego system has been extended twice, and there are plans to create an area-wide network (Figure 7.2), even though costs have risen due to inflation and building more difficult routes. The El Cajon line cost $9 million per mile, the Santee line $34 million per mile, the Bayside $37 million, and the Mission Valley (West) $36 million. Future plans include the Mission Valley line at $55 million per mile (Figure 7.3), which includes some tunneling, and the Mid Coast line at $32 million.

Figure 7.2 San Diego light rail network (Map: San Diego Transit)

Figure 7.3 Orange line (Mission Valley East) low-floor cars (Photo: San Diego Transit)

These costs are very frugal compared to the recent systems built in the UK at $50 million per mile at grade or in Buffalo, mostly in tunnels (5.4 out of 6.4 miles), at more than $100 million per mile. While the Buffalo system has not reversed the long-term population decline of the city, households near stations have seen some benefit. A study by a University of Buffalo urban planning researcher found that houses located within a half-mile (800-m) radius of Buffalo's light rail stations are assessed at $1300 to $3000 more than similar properties not within walking distance of the stations.

No property value effects have been reported in San Diego, and the system's continuous development has been justified by its success in attracting car trips from congested freeways. Making the core urban centers more accessible and reducing the need for car parking in those centers are other benefits of light rail.

The Transportation Research Board (1992) noted that in San Diego, "although the basic philosophy of low cost, high speed, and primarily at-grade is still the foundation of the design process, 10 years of design experience, together with changing socioeconomic conditions, have resulted in the design approach being modified to meet the needs of a rapidly expanding [light rail transit] system. Experience has taught lessons that have been incorporated in subsequent design efforts." Sharing such lessons is important to ensure that new light rail systems are economic to build and operate and maximize the diversion of car trips to reduce oil use and polluting emissions.

7.2 Calgary, Canada

Hosting the Olympic Winter Games spurred the installation of a light rail system in Calgary. Calgary was strongly influenced by San Diego in choosing standard vehicles and as simple an operating system as possible. The new light rail line in Calgary also opened in 1981, using the same Düwag U2 light rail vehicles as San Diego. Calgary and San Diego together set the fashion for many of the subsequent light rail lines built in North America over the next 25 years.

The Calgary light rail transit (LRT) is now 49 km (30 miles) long in two lines with a fleet of 82 U2 and 32 SD160 cars (Figure 7.4), serving 37 stations, including those in the free zone downtown area. The daily ridership in 2008 was 297,500, about 94 million yearly. This compares to 285,000 daily riders on the more extensive Toronto Streetcar network. Although the population density of Calgary is not high, like many North American cities, the high ridership can be partly attributed to the strategic park-and-ride, kiss-and-ride, and bus-and-ride stations. These greatly increase the effective catchment area of the system. Calgary presently has a network density of 3.1 million riders per mile of network.

Figure 7.4 Calgary LRT car (Photo: Calgary Transit)

Calgary is also proud of the cost effectiveness of the construction of the LRT network (Table 7.1). Some of the other lessons learned in Calgary (Hubbell and Colquhoun 2006) are equally applicable to other cities considering light rail:

- Build the right *project*: Ridership and cost estimates must be "rock solid" and the most appropriate technology must be chosen.
- Build the project *right*: Exploit the design flexibility of LRT to use surface alignments.
- Protect future land requirements: Plan for the future and secure land in the present.
- Develop the transit corridors: Use existing modes to create the corridors that can be exploited by LRT
- Use the right strategic operating practice: The skill sets and interests of planners, designers, and builders are fundamentally different from the operators and maintainers. There should be separate positions for LRT operators. Minimize maintenance and system expansion during nonrevenue hours.

Table 7.1 Reported unit construction costs in Calgary

Item	*Canadian $ million*
Elevated track per kilometer	30
Tunneled track per kilometer	35
At-grade track per kilometer	15
Station	2

- Find the right *champions*: Strong and continuing support is needed from key officials and politicians, as well as a good body of stakeholders, including the local media.

Since the Calgary LRT was opened in 1981, the number of jobs in the central business district has grown by 18,000, with a reduction of parking provision, no increase in road capacity, and an increase in mode share by transit to 45% of all downtown trips. These are important benefits, of which only the increase in LRT ridership ends up in the annual transit accounts. The Calgary LRT also has helped to concentrate suburban job growth in employment centers at LRT stations, thus encouraging less driving even in the suburbs.

The Calgary LRT also has reduced accidents (Table 7.2) while at the same time increasing ridership. Part of the reason for the low accident rate per boarding LRT passengers has been the attention to good station design. The main principles are:

- Short access time to station with an at-grade approach
- Integrate the station with its surroundings
- Use the highest "barrier-free" accessibility design standards
- Stations should be user friendly and be a significant local community facility
- Have different station entrance positions to even light rail vehicle loading
- Create a high level of personal security by "defensible space"

The high level of ridership can be attributed to the integration of mode access policies. This has meant that as far as possible every station should be accessible on foot and by bike, car, taxi, and bus. In addition to park-and-ride to divert longer distance commuters, key LRT stations also benefit from

Table 7.2 LRT and bus accident rates compared

	Mode	
Item	*LRT*	*Bus*
Collisions per million kilometers		
1995	11.3	23.0
2005	10.3	17.8
Accidents per million passengers		
1995	0.4	5.6
2005	0.06	1.6

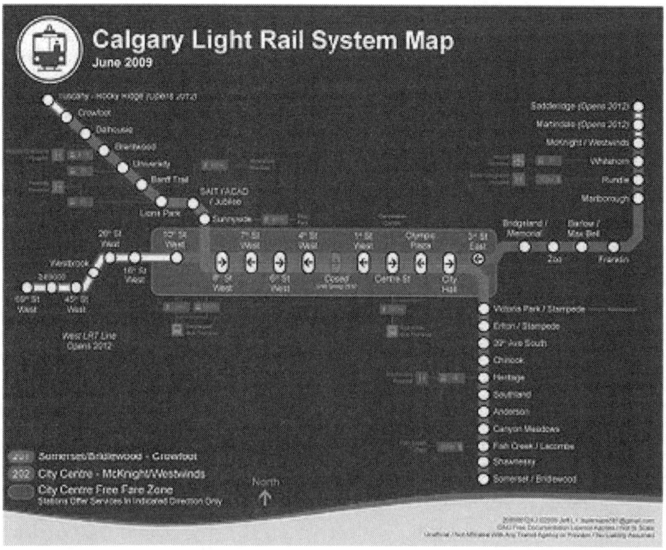

Figure 7.5 Calgary LRT network (Map: Calgary Transit)

feeder bus routes. In Calgary, this was easy, as bus and LRT are provided by the same organization. In other cities, where such internal cooperation is rarer or bus and LRT are operated by different organizations, then the LRT must provide inducements to get cooperation. One inducement might be simply to show that the bus operator is better running on free-flow roads, feeding LRT, than being on congested roads into the city center.

There are ambitious plans to extend the light rail network in Calgary (Figure 7.5). Route 202 will be extended westward from the city center, new lines 203 will run southeastward and 204 northward, and 205 will branch off 202 to serve Calgary International Airport. Calgary light rail vehicles are powered by wind generation fed into the local grid and supplied to LRT substations. In this case, the LRT is both fossil-fuel free and does not emit any carbon dioxide. Indeed, if the car trips attracted to the LRT are included, Calgary makes an impressive impact on reducing local carbon dioxide emissions.

This expansion is built on the solid foundation of success in Calgary, where the LRT has reversed the decline in transit use, reduced car traffic, increased transit mode share, and most importantly shown that LRT can successfully and safely use alignments in city streets. In addition, LRT enjoys priority over other traffic, so it always has consistent journey times without suffering from congestion.

In addition to these concrete benefits from LRT, it is consistently at the top of the list of citizen satisfaction surveys conducted by the city over the last decade. Keeping the quality of service high, to maintain this level of satisfaction and ridership, will keep the management of Calgary Transit on its toes.

7.3 Karlsruhe, Germany

In Europe, Germany deserves much of the credit for maintaining urban rail transit and developing light rail (technologies). After the Second World War, oil, rubber, and labor shortages meant that tramways were the only affordable mass city transport in the 1950s. Later, new higher capacity articulated vehicles were perfected by Düwag in the 1960s. Then high-speed light rail vehicles were developed in the 1970s for large cities like Frankfurt am Main. These U cars became the backbone of new North American systems in San Diego and Calgary starting in 1981. Then low-floor light rail vehicles were developed to allow level boarding from low platforms in city streets. This gave an impetus to building new systems, as disabled access could be offered at an affordable cost.

The prize for the most radical and far-reaching development however must go to Karlsruhe. Here, two practical problems collided and produced a convenient and economic solution. The railway station in Karlsruhe is a long walk from the city center. The local railway network had a number of loss-making branch and rural lines with infrequent service. The technical solution was light rail vehicles able to operate from overhead lines at both the tramway voltage (750 DC) and main line railway voltage (15 kV AC). The main line operator, Deutsche Bahn, was happy to transfer local lines to the city operator (Karlsruhe), which allowed Deutsche Bahn to concentrate on the improvement of intercity service. Light rail vehicles running at higher and regular frequency replaced sparse trains. A light rail link was built to allow light rail vehicles to run from rural termini over street tracks (Figure 7.6) to the city center and other destinations.

The result of faster and frequent light rail vehicle service was a significant growth in traffic, much of it captured from congested roads. With lower operating costs and higher revenue, the branch line economics were reversed, with a virtuous growth cycle. This process of converting urban and rural branch lines into light rail integrated in an urban network has been very popular. The Greater Karlsruhe network is now 400 km long (Figure 7.7), compared to just 80 km in 1980.

How were the risks of interoperating heavy rail trains and light rail vehicles managed? This was achieved by strict signaling protocols and scrupulous time keeping. There have been no collisions. The idea of a tram running on a railway line has been named "tram-train." The main technical innova-

Figure 7.6 Kaiserstrasse and Markplatz hub (Photo: Karlsruhe Transport)

tions include hybrid wheel tire profiles to accommodate railway switches and points, and track management and signaling procedures. There are now tram-trains operating permanently in Saarbrucken (international Dutch/German service). Tram-train trials in Ottawa led to a decision to convert busways to light rail. In the UK, however, tram-train has not progressed beyond debate.

In terms of providing seamless travel from suburbs and rural areas into central city destinations, tram-train is a genuine integrated service. The many "before-and-after" studies indicate that tram-trains are popular with residents, who can leave their cars at home or convenient park-and-ride and tram-train to work, shopping, school, etc.

The economics of urban and rural branch line railways can be replicated in many countries. The 1981 US Railroad Deregulation (Staggers) Act allowed Class 1 railroads to pass the operation and ownership of "short lines" to local business and municipal consortia. This is the best of both worlds, as feeder branch line traffic is retained and developed by enthusiastic local companies. Adding light rail, as San Diego did, further improves the finances and the economic benefits.

Karlsruhe began a revolution in the way urban passenger rail services can be operated and funded. This shows that the ownership and operation of rail are not cast in stone. With modest investment (e.g., for overhead line electrification) and a willingness to manage in a different way, more car commuters can have access to sustainable light rail. This, of course, was a lesson learned from Skokie Swift in 1963, forgotten from the long-lost "interurban" closed in the 1930s and reinvented in Karlsruhe in 1981.

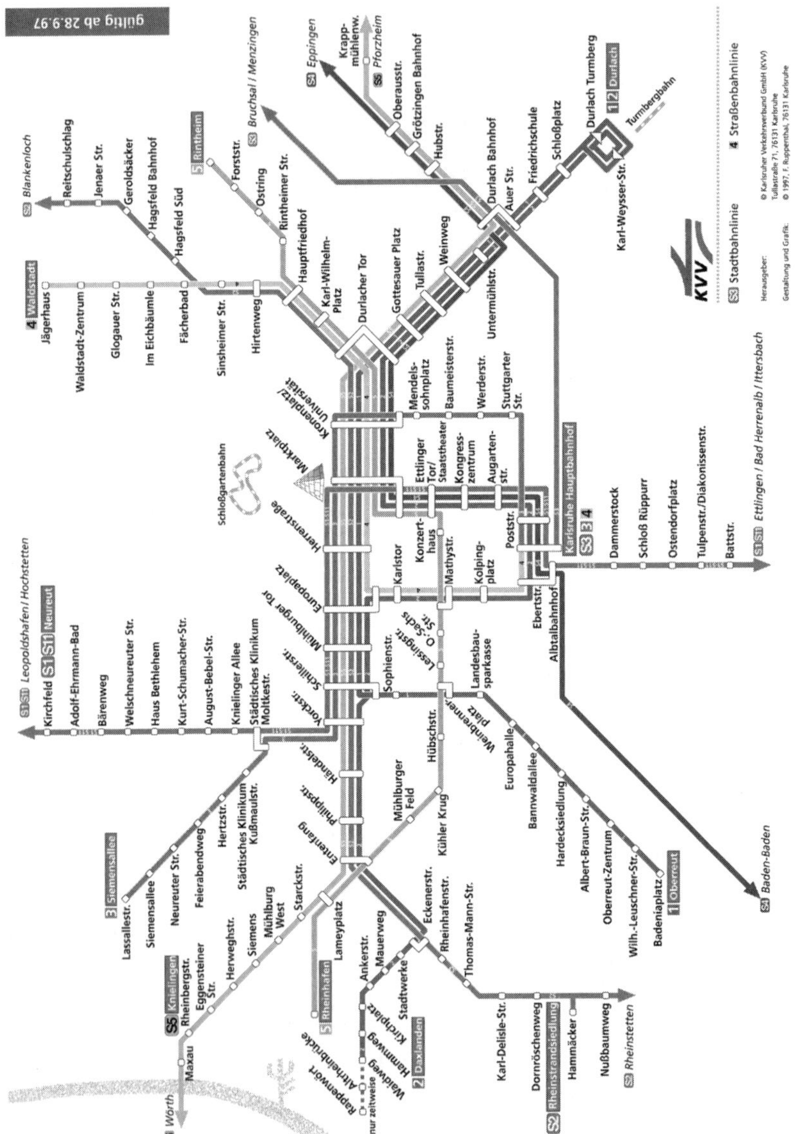

Figure 7.7 Karlsruhe light rail vehicle network (Diagram: Karlsruhe Transport)

7.4 Nantes, France

Nantes was the first new-generation tramway in France, which like the UK and much of North America had abandoned its original tramways in the 1950s and 1960s. The new tramway opened in 1985, and the first results were reported at a conference in England (Pitrel 1987). The total cost of building and equipping this new light rail line was FF630 million (about US$100 million) for the first line, some 10.6 km long, or about $10 million per kilometer.

Nantes has a metropolitan area population of about 408,000, and although the original tramway was electrified in 1913, it was progressively abandoned after 1930, with the last line closing in 1958. The principal reasons for the demise of this tramway were lack of political interest and investments in improving the road network. After the oil crisis of 1974, the French government introduced a policy to reduce dependence on (imported) oil and encouraged local authorities to seek ways to achieve that by offering 50% grants toward the capital cost of new light rail systems.

The first light rail line opened with 20 articulated high-floor light rail vehicles, 28.5 m (94 ft) long, on 6 axles (Figure 7.8), offering 58 seats and 112 standing spaces. The light rail vehicles operated in single-car sets every 5 minutes during the peak, 7.5 minutes between the peaks, and 10+ minutes at other times. This line immediately captured 20% of all transit trips in

Figure 7.8 Nantes original two-car high-floor and added low-floor center (Photo: Nantes Tramway Authority)

Table 7.3 Passenger access to Nantes LRT

Access mode	*% of trips*
Walk	64
Bus (transfer)	27
Park-and-ride/kiss-and-ride	5
Other	4

Table 7.4 Before-and-after study data

	Year		
Trips (million)	*1984*	*1985*	*1986*
By bus	51	50	53
By LRT	0	8.2	11.9
Per vehicle kilometer			
Bus	3.3	3.8	4.3
LRT	0	11.8	13.1
Rides per capita	107	123	137

Nantes. Access to the new light rail system shortly after opening was surveyed (Table 7.3).

A before-and-after study was undertaken to determine the impact of the new light rail system in Nantes (Table 7.4). The last full year (1984) without LRT, the opening year (1985), and the first full year of operation (1986) were compared.

The before-and-after study also looked at the impact of the new tramway on other traffic. Traffic speeds were unaffected by the priority afforded to LRT. Parking demand in the city center was reduced with the mode switch from car to light rail vehicle. The immediate success of the Nantes system was such that extra light rail vehicles were purchased to allow two-car consists to operate during the peak period to relieve overcrowding.

The continual growth of LRT traffic led to line 2 being opened in 1992 and line 3 in 2000, by which time light rail vehicles were carrying 40% of all transit trips in the city. The network length is 41.4 km (26 miles), requiring 79 light rail vehicles to serve the 90 stops. By 2007, a total of 63.5 million light rail vehicle trips were recorded (Table 7.5).

During this period, there also was modest growth in bus patronage, running on a smaller network. The full cost of bus and LRT operation is shown in Table 7.6. Here, LRT has a lower cost per vehicle kilometer, through lower fuel and maintenance costs. What is more illuminating is that the LRT covers

Table 7.5 Nantes LRT ridership

Line	Trips (million) 2007
1	27.6
2	24.1
3	11.8

Table 7.6 Operating costs and revenue

	$/vehicle kilometer	Revenue/cost
Bus	2.6	48%
LRT	3.6	113%

its operating costs from passenger revenue, whereas the buses need to be subsidized.

LRT ridership growth continued to 2008, when a daily average ridership of 256,000 was recorded, equivalent to more than 75 million trips annually. In planning the LRT lines, great care was taken to serve the active parts of the city. Of the total employment in the city, some 40% of all jobs are within walking distance of LRT stops. Similarly, 80% of schoolchildren can use the LRT to travel to and from school. What Nantes has demonstrated is that a well-planned LRT system can capture a significant market share from car trips and in doing so improve the financial performance of transit.

7.5 Sheffield, England

Sheffield closed its first tramway in 1960, 2 years after Nantes. Old trams were replaced by new diesel buses. Although the logic behind the closure was the high cost of replacing worn track and extending the network past the 19th century city limits, it soon was clear that closure was a mistake. Ridership plunged and traffic congestion worsened, but the political embarrassment of the city council meant it would be another 35 years before trams ran again in Sheffield. There are pictures of some of the Sheffield light rail vehicles in various liveries throughout this book; this section will focus on the buildup to the (re)opening of the tramway and the impact it has had on the city subsequently.

During the 1970s and 1980s, the transit authority, known as South Yorkshire Passenger Transport Executive (SYPTE), attempted to make the best of an all-bus system by implementing a low-fare policy, freezing fares at the 1975 level. On this basis, with inflation, the real cost of tickets would decline.

It was expected that by 1985 fares would be less than the cost of collection. At this point, it was planned that fares would be abolished and public transport would be free (but paid from taxes). A change of UK government in 1979 made that impossible politically and the SYPTE suffered severe financial penalties. Therefore, in 1986 bus fares were raised by 300%.

The low-fare policy stabilized bus use compared to the rest of the UK, where bus use continued to decline. There was, however, no evidence that car users were attracted to make more or any trips by bus. "Captive" riders, however, did make more frequent trips, which compensated for those buying cars and abandoning transit. This was researched in some depth by the Transport and Road Research Laboratory (Hay 1986a, 1986b), which found that every first car bought by a household reduced bus use by about 390 trips per year.

When fares were raised, this served as an incentive to families that had not bought their first car to do so. Ridership collapsed from about 200 million bus trips a year to about 100 million. Car ownership and usage soon caught up with national average levels.

New roads built in the previous decade did not have enough capacity, so the city and transit authority began a new round of studies to solve the problem. One option reconsidered in depth was personal rapid transit (Cabtrack) (Langdon 1971). A detailed case study included photomontages of the elevated track through different parts of the city (Latour 1971). Personal rapid transit was discarded as not having enough capacity and costing too much (Lowson 1999).

Against a background of rising car ownership, use, and congestion and declining bus patronage, SYPTE realized the something other than buses was needed to attract car users in order to reduce congestion. After several studies, three corridors were identified as having the highest demand and the potential for further traffic growth. These corridors were worked up to become the core lines of the Supertram system (Figure 7.9), a 29-km- (18-mile-) long network of three lines, meeting in the city center of Sheffield. The cost when opened in 1995 was £240 million (about $380 million), for a cost per kilometer of £8.3 million (about $21 million per mile).

There are 48 stations, and the system operates with 25 three-section light rail vehicles built to SYPTE specifications by Düwag (Siemens). These light rail vehicles are 35 m (115 ft) long and weigh 52 tonnes, the biggest single-unit light rail vehicles at that time. The vehicles had 88 seats when delivered, which was reduced to 86 after refurbishment. There also are 154 standing spaces and 2 wheelchair spaces. The vehicles also enjoy 100% motored axles to allow the 10% maximum gradients to be climbed. Steeper grades are found in Lisbon (12%) and previously in Saarbrucken (13%) and Montreal (14%).

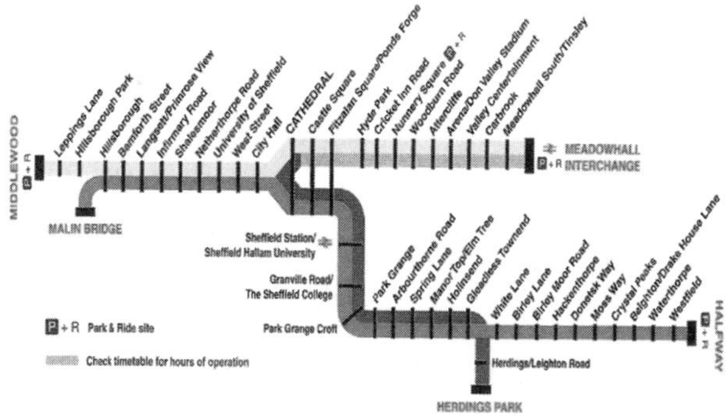

Figure 7.9 Sheffield Supertram map (Map: South Yorkshire Passenger Transport Executive)

No further versions of this light rail vehicle design of mixed low and high floor were made.

On opening, some problems were encountered, not all of which have been addressed 15 years later. In particular, the operating speed was slower than that of contemporary continental tramways. There are several reasons for this (Bygate 1995):

- Delays at traffic signals without preemption
- Congestion in mixed street operation
- Slow passenger boarding/alighting at stops
- Operational and technical faults

Due to budget constraints, the system opened without transport integration, such as:

1. Feeder buses
2. Park-and-ride/kiss-and-ride
3. Cycle parking
4. Easy pedestrian access
5. Through/interchange tickets

This meant that in 1996, the first year of full operation, ridership was only 6 million, about half that forecast (National Audit Office 2004), although the forecast did assume a period during which patronage would build up to the final level. Since opening, some of these issues have been addressed, but there are others that have not yet been resolved which helped to make the system initially operate at a loss.

All 48 stations are low height (375 mm/15 in. to the top of the rail). Some of the stations on street running sections are integrated with the footways at the side of the carriageways. Others on reserved track sections are more like conventional railway stations and have similar restricted access.

Ridership in 2005/2006 was 13.15 million per year, with about 36,000 trips per day. In 2008/2009, ridership rose to 15 million, requiring a staff of 250 to operate and maintain the system. This growth has been achieved after the system was privatized in 1998 to the multinational Stagecoach Group for £1.25 million with an operating concession until 2024. A recently appointed new general manager, Andy Morris, was a breath of fresh air after the municipal approach (Jones 2007) and led the refurbishment of the fleet (Figure 7.10). Stagecoach also concentrated on the containment of operating costs, and the growth in revenue means that the Supertramway makes a surplus of about £1 million a year. This is an excellent return on Stagecoach's 1998 investment. The opening of two new stations to serve redevelopment sites also helped to attract more traffic and revenue.

Figure 7.10 Supertram in new livery (Photo: L. Lesley)

7.6 Sydney, Australia

Like the other case studies reviewed in this book, Sydney had a first-generation tramway; at its maximum (290 km/180 miles), it was the biggest network in Australia, with the steepest gradient at 12%. This large network required almost 1600 cars and carried 405 million passengers at its peak in 1945. This intense service required a tram every 20 seconds to cross the junction of George and King Streets. Also, like the other case studies reviewed, it was progressively abandoned beginning in 1939 until the last tram ran in 1961 (Keenan 1979).

The replacement bus service operated at a loss from the outset, unlike the profitable tramway. The city council regretted the closure within a couple of years, but the abandonment had been emphatic, difficult, and costly to reverse, with the overhead lines taken down immediately and the tracks covered over.

This did not prevent proposals being made in 1975 and then in 1995 for a tram loop between Central Station and Circular Quay, which came to nothing, although a one-way monorail loop was installed in 1988 around the city center as a tourist attraction and as part of the redevelopment of land at Darling Harbour (Churchman 1995) (Figure 7.11). In terms of a transit mode, the monorail has a limited role in Sydney and has had three separate owners in its 20-year life, indicating difficult operating finances.

The Sydney bid for the 2000 Olympic Games provided a catalyst for another attempt to improve the transit facilities of the city and in particular serve the Olympic Village with a high-capacity high-performance system. Sydney Light Rail Company, a private venture, was formed in 1994 to promote, build, and operate a new light rail system. Construction began in 1996 and the system opened for service in 1998, 16 months later. Construction was accelerated by the use of former tramway and railway alignments. The

Figure 7.11 Sydney monorail (Photo: Sydney Transport Authority)

total capital cost was A$85 million, about US$68 million, which works out to US$15 million per mile.

Connex was appointed as the operating company, and it also runs the monorail service. The light rail line is 7.2 km (4.5 miles) long, has 14 stations, and requires 7 light rail vehicles to provide a 6-minute-frequency peak service. The light rail vehicles are a low-floor design, with the entrance 300 mm (12 in.) to the top of the rail. Each vehicle costs A$25 million, about US$2.9 million. The "Sydney" Bombardier Variotrams are 28 m (92 ft) long, with 74 seats and space for 102 standing passengers. They weigh 36 tonnes and have a maximum speed of 70 km/hour (44 mph) (Figure 7.12).

For many Sydney residents and tourists, the light rail line has been a disappointment because of the lack of geographical coverage. There are other transport corridors with larger traffic potential. Unfortunately, differences between the city council and government of New South Wales have meant that funding for extensions has not been forthcoming.

Patronage has been less than forecast partly due to the imposition of a slow speed for the street sections of the line, making journeys longer than by car or bus. Nevertheless, there has been a 15% reduction in cars entering the central business district and a reduction in the emission of carbon dioxide. Planned extensions, when funded, will seek to overcome some of the design issues from the first phase. Adding further light rail lines also will increase ridership through network synergy. Some of the issues being considered for future extensions are:

- Improving integration with other modes
- More park-and-ride stations to capture car commuter trips

Figure 7.12 Variotram in Sydney (Photo: Sydney Transport Authority)

- Joint ticketing to make through trips easier
- Better station design
- Better tram design

Like many cities faced with high-cost tramways, a significant extension of the Sydney system seems unlikely until ways to reduce costs can be found, even in light of the impact of environmental and oil issues.

7.7 Galway, Ireland

Galway had a single-horse tram line that operated between 1887 and 1924, when it was abandoned without electrification. In contrast with the other case studies above and most other new light rail lines opened in the last 25 years, Galway is a privately promoted and funded initiative. While the international credit crunch has eroded the public sector's ability to fund new transportation infrastructure projects, the need for light rail is growing. New light rail systems respond to climate change challenges, reduce carbon and health-threatening emissions, and reduce dependence on petroleum.

The Irish city of Galway has a population of 85,000 and is facing this dilemma in a different way. The business community sees daily and severe traffic congestion hampering the economic strength of the city. Most people live on the west side of the city and much of the employment is on the east side of the city, with only four bridges across the River Corrib, which splits the city in two. GLUAS is an Irish word meaning "movement" and has been adopted as the brand name for the new light rail system that is being constructed.

Presently only 4% of all trips are made by transit (buses) and 65% are made by private car, with traffic volume growing. The GLUAS4Galway Group commissioned a feasibility study to identify routes that are physically possible to be used for light rail. This study included abandoned and little used railway lines. In this exercise, a first-cut capital cost was determined for the seven routes identified.

From this, an economic evaluation was undertaken. The likely ridership that each of the seven lines could expect was calculated. Generalized costs were used to determine the number of trips based on the resident catchment populations, quality of service proposed, and likely fare levels. This evaluation ranked the viability of each route in terms of the likely operating surplus compared to the capital cost using the internal rate of return (IRR).

The routes with the best IRR were then combined into a network of two lines, 21 km long (Figure 7.13). While this network had a positive IRR of about 7%, the GLUAS4Galway Group undertook further work to improve

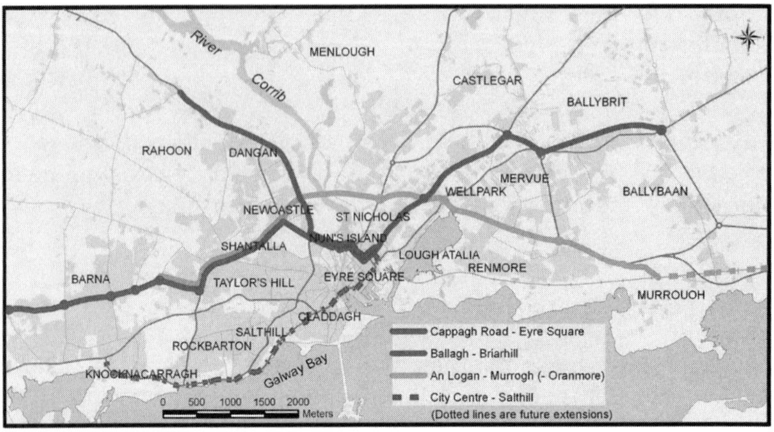

Figure 7.13 Planned GLUAS light rail system (Drawing: GLUAS4Galway Group)

the financial performance. An IRR of 7% was not considered large enough to attract private funding (Table 7.7). A decision was made to consider embedded renewable power generation in light rail investment.

After considering the power requirements of the light rail system (maximum demand of about 9 MW) and the variability of generation, it was agreed that a 2-MW wind generator would be installed at each of the five planned park-and-ride stations and 4 MW of hydroelectric power would be generated on the River Corrib. The average power demand of the Galway light rail system is 2.5 MW. Surplus power will be sold to the national grid, for which there is a guaranteed buy-in price. Not having to purchase power and selling a surplus improved the IRR to more than 10%, considered to be the minimum needed to be commercially viable (Table 7.8). Since that exercise was undertaken, the guaranteed buy-in price was raised by 10%, raising the IRR of the GLUAS network from 11.3 to 13% a year and making it financially

Table 7.7 GLUAS base case

Route	*Capital cost (€ million)*	*Operating cost (€ million)*	*Passengers per year*	*Revenue (€ million)*	*IRR (%)*
1	56	1.85	801,000	2.40	0.99
2	41	1.35	1,412,000	4.23	7.10
3	38	1.12	763,000	2.29	3.09
4	35	0.61	227,000	0.68	0.19
5	31	1.09	708,000	2.12	3.32
6	35	1.15	960,000	2.88	5.00
7	37	1.26	545,000	1.63	1.01

Table 7.8 GLUAS with embedded renewable generation

Route	Capital cost (€ million)	Operating cost (€ million)	Passengers per year	Revenue (€ million)	IRR (%)
1	57	1.85	801,000	4.29	4.3
2	42	1.35	1,412,000	5.12	11.4
3	39	1.12	763,000	4.18	7.9
4	36	0.61	227,000	2.15	5.4
5	32	1.09	708,000	4.01	9.0
6	36	1.15	960,000	4.77	10.1
7	38	1.26	545,000	3.52	6.0

viable. Finally, a business plan showed that the cost of this light rail system would be €200 million and would need 17 light rail vehicles operating at a 6-minute frequency to carry 1.5 million passengers in the first year, which was projected to grow to nearly 8 million in the fifth year.

Data from Table 7.8 was used to identify viable networks using different combinations of the original seven light rail routes evaluated. The results are shown in Table 7.9.

GLUAS4Galway set up a website in June 2008 (www.GLUAS.com) and in September 2008 made a presentation to the city council and got unanimous support for the project. The city council appointed MVA transportation consultants to review the GLUAS project. This confirmed in 2009 that the capital cost of €200 million was achievable. A joint working party is now considering the mechanics of installing a light rail system through the historic core of Galway (Figure 7.14).

Part of the insight of the GLUAS4Galway Group was recognizing the need to create investor confidence. This was achieved by the use of embedded renewable power generation, which means that even before the first light rail vehicle runs, revenue flows from sales to the national power grid. Each of the five park-and-ride stations will have a wind turbine. This will both provide power for the light rail vehicles and also act as a landmark for commuters driving into the city. Hydroelectric generation will be installed on the River

Table 7.9 GLUAS network evaluation

Network	Capital cost (€ million)	Operating cost (€ million)	Passengers per year	Revenue (€ million)	IRR (%)
A = 2 + 3, 6, 7	150	4.00	4,538,000	16.10	10.4
B = 1 + 3, 5, 6	159	4.10	3,883,000	14.66	8.8
C = 2 + 3, 5, 6	144	3.84	4,817,000	16.89	11.3
D = 2 + 3, 4, 6, 7	185	4.77	3,180,000	10.05	5.0

Figure 7.14 18th century bridge, cathedral, and historic core (Photo: L. Lesley)

Corrib. By selling the surplus power to the national grid, GLUAS will have a dependable income stream from day one. Income will increase as patronage and passenger revenues build.

The key to making GLUAS affordable as a commercial investment has been the choice of equipment and only building what is needed for a tramway. Where the city council seeks civic amenity improvements, these will be undertaken on a partnership basis. Presently, the assessment is that 17 light rail vehicles will be needed to offer a peak service frequency of 6 minutes on each line. The cost assessment is based on the use of the City Class tram, which has a 200-passenger capacity with nearly 100 seats and is the most power-efficient light rail vehicle on the market, with an average consumption of 1 kWh/km run. This is about a third of what other products use and means that in about 10 years of service, power savings will equal the purchase price.

The capital costs will be raised from a mixture of loans and the sale of shares, about 50% from each. GLUAS will operate at a surplus. Sensitivity analyses were undertaken to determine the impact of competition from buses, lower fuel/parking costs, and higher operating costs. These all would reduce the operating surplus, but not significantly. Investors know that the surplus is large enough to pay dividends and to retire the capital debt.

Figure 7.15 Trans Lohr system in Clermont, France (Photo: Brendan Holland, GLUAS4Galway)

The MVA study also looked at bus competition and predicted that a "perfect bus network," with zero waiting time, would attract a maximum of 8% of all trips in Galway compared to the existing 4%. In comparison, the two-line GLUAS network planned will attract 16% of all trips and thus increase the total public transport (transit) market share to 20%. Further, GLUAS has potential for increased growth. New lines will be considered, as well as extensions to existing lines. With an established network, these extensions can be costed on a marginal basis. The GLUAS ridership forecasts exclude all external trips coming into the city in the morning that are attracted to the five strategic park-and-ride stations.

The Galway light rail system will thus be environmentally, energy, and financially sustainable—possibly the first in the developed world in the 21st century. The promoters of GLUAS4Galway hope it will provide a model for other light rail systems in many smaller towns and cities in Europe and North America.

Finally, Trans Lohr (Figure 7.15), which markets a guided bus system, visited Galway to offer its system instead of "old-fashioned trams." Its assessment of the capital cost of the same network was €400 million. Recently, the operations manager of the first Trans Lohr system in France, speaking at a Union Internationale des Transportes Publiques (UITP) Conference, stated that the maintenance costs are much higher than a tramway and the whole life costs of a tramway are lower.

8

Conclusions

This book has reviewed a considerable body of experience and research in planning, designing, engineering, building, and operating new light rail systems. It shows that more than 100 years of experience can be drawn upon to find cost-effective solutions to the problems of installing a new system in existing urban areas. There is no doubt that a new light rail system will provide considerable transport, environmental, business, and community benefits.

Two models have been examined. In the first, public authorities use public money to generate wider benefits. Generally, public systems tend to be in regulated markets with fares set politically. In the second, private companies use private funding to build light rail systems that produce operating surpluses, which can service the capital debt of the private investment. The private funding model is most likely to succeed where regulation relates to health and safety matters with the commercial freedom to set fares that reflect market conditions.

8.1 Meeting Community Needs

People need access to work, schools, shops, hospitals, leisure activities, and the other important functions of society. Clearly, this cannot be achieved by everyone driving private cars everywhere. In freeway-rich North American cities (Figure 8.1), this is not possible, as attested daily by serious traffic congestion and air pollution. In high-density cities, like those in Europe, it is only physically possible to satisfy a very small part of the demand to use private cars. This has been known since at least 1963 (Buchanan 1963).

Figure 8.1 Freeway land use (Photo: Los Angeles County Metro)

Even in the fully motorized society of the US, the land needed for more freeways and parking consumes a significant part of an urban area. Economists demonstrate that land used for roads and parking becomes nonproductive and reduces the economic output of the city. There is a critical density of such economically "unproductive" land.

This ignores the energy issues such as dependency on oil and the environmental problems from greenhouse gas leading to global warming and other health-threatening pollution (Figure 8.2). Finally, there is no level of urban road building that can avoid peak-period congestion, which in Los Angeles now lasts from before 7:00 a.m. to after 7:00 p.m. on working days in spite of having a freeway system bigger than the entire UK (which has a population of 60 million).

Having access to the services people require without needing to use cars means being able to walk or cycle for short journeys (in the UK, 50% of all trips are under 5 km [3 miles] long) and public transport, which includes taxis for longer trips. In many cities, public transport means a bus network. Data shows that bus use has fallen as car ownership has risen and that car users do not find buses a satisfactory alternative to their cars.

The advantage of light rail is that it is proven to be an acceptable alternative to private car use and that car users will park and ride from a light

Figure 8.2 Traffic smog (Photo: Los Angeles County Metro)

rail station. In addition, electric light rail emits no pollution in city streets and is quiet, comfortable, and accessible to all groups of people in the community. With renewable electricity generation as in Calgary and Galway, light rail also can be completely energy sustainable. Using renewable power helps to reduce greenhouse gas emissions and the need to import oil. This is a double benefit when diverted car trips that save oil and reduced greenhouse gas emissions are included.

8.2 Satisfying Market Demands

Many urban areas have travel patterns many-to-many in nature. Many city centers/central business districts (CBDs) have lost their dominant economic role. In 1960, some 70% of all trips in Liverpool, England were to or from the CBD. By 2000, this had fallen to 30%. A new light rail system must cater to these non-CBD-based trips and, from experience with new lines in the US, also help to re-establish the role of the CBD by releasing car parking (in Liverpool, 70% of CBD work trips are by car). Changing the modal split for CBD trips will enable more intensive economic activities to be developed. These can be accessed by more people using a light rail system and reduce the need for CBD car parking.

With the growth of e-trade, will the need for face-to-face contact also be reduced? Previous experience shows that better communication stimulates travel. The 10 years of e-communication has allowed an even wider group of people to make contact, but has not reduced the need for face-to-face meetings. Indeed, such changing markets can stimulate use of socially acceptable light rail by raising awareness of the services available.

While the way we work and earn our living will change, as will shopping, education, leisure activities, etc., the basic need of human society is personal contact. New light rail systems will enable more people to make contact and travel with less resource consumption and less environmental damage. New light rail operators will use the Internet to find out the travel needs of residents in the area and tell them about the convenience and advantages of the services offered. Light rail should become a cashless activity and thus be perceived to cost less than driving a car for all trips.

There could be a time when light rail systems offer fuel coupons for regular users when fuel is $15 a gallon, as a special incentive, or to reach areas not directly served by light rail. Similarly, a light rail ticket might include a taxi link or a car service. Finally, bike-and-ride with cycle trails to light rail stations is another option for future urban travelers, already enjoyed by residents in many European cities.

8.3 Achieving Commercial Viability

There is no doubt that public funding is under strain in many countries. Finding the money to construct a new light rail system is however a one-off expenditure. If, on the other hand, the system does not generate enough revenue from its passengers to cover the operating costs, then there will be a permanent demand for subsidies. It is therefore important that however the capital costs are met, the operating costs should be covered by the passenger revenue. Indeed, there should be enough surplus so that a depreciation fund can be built up to pay for midlife refurbishment and end-of-life replacement of assets. This is the minimum for commercial viability.

Chapter 5 set out some of the ground rules for that and techniques for ensuring that at the planning stage the promoter can be confident that the new light rail system, when open, will be commercially viable. Chapter 6 showed how once open, the operator can ensure that the maximum patronage or revenue can be attracted by appropriate pricing and good advertising.

8.4 Learning from Experience

There are nearly 500 light rail systems operating around the world, some with continuous operation for over 100 years. There is plenty of opportunity both to learn from earlier experiences and to take parties of decision makers to see and sample what a light rail system is and how it works. This practical contact can be an important step in achieving a new system. A light rail promoter cannot assume that the people in the area will know or understand what is being proposed. Indeed, some may imagine that a railroad is being planned outside their front yard. This is particularly true if the new system will lead to the demolition of property.

Here, decision making may change from rational to emotional, especially if people's homes are involved or an old cultural building is being threatened. The promoter should therefore seek routes and light rail alignments that can be achieved without the need for demolition. In addition to reducing the political issues, this will speed up the implementation and minimize the cost. Chapter 7 presents some case studies from around the world where problems have arisen after opening but have been overcome, often using creative or innovative solutions.

8.5 Get It Right the First Time

All investors in light rail, whether the community via its taxes or private financiers, will seek the best value for their money from the new system. Learning from experience should help avoid earlier mistakes. Just as important, though, is the need for an integrated design, construction, and operating team, so that before the system is built, potential problems can be identified and solutions arrived at that will satisfy all parties. The contractor who will construct and install the light rail system often can propose easier ways of undertaking the project. The operator will have in mind the need to avoid problems that could increase costs or make the service less convenient every day.

Chapter 4 sets out the engineering issues of converting a plan into reality. In particular, constructing a light rail system through the center of an urban area is very different from a self-contained building site with a hoarding around it, where work is hidden from public view. The funders, politicians, and the public will be able to witness the progress of the light rail system. Good public relations, with an information campaign, can let people know what is happening and why. This can at least minimize misconceptions. Keeping the local press on board is also important, since bad news sells papers. Therefore, building the system once and opening it promptly is the best policy for a successful light rail promoter and should lay the foundation for system expansion as people see the benefit as riders and those still driving see improved traffic conditions. As Mogridge (1990) pointed out, the best way to reduce road traffic and congestion is to improve the journey speed of competitor public transit and attract marginal car commuters off the road. Light rail is an affordable and proven way to do just that.

8.6 Diversification

A light rail, or any other transit, system is not an end in itself. It is the means whereby an urban area can function efficiently and economically. It allows people to get to work or school, do their shopping, and enjoy leisure activities and social contact. While the light rail operator should be very efficient in providing a safe and convenient service, there also needs to be links with the activities that the system serves. In particular, joint ventures with attractions can offer combined tickets at an attractive price and allow light rail customers to enjoy a popular attraction at a discount price.

Similarly, seasonal events (e.g., St. Patrick's Day, Christmas, etc.) can be used to develop themed material to attract riders who might otherwise not use transit. This is especially important if there are events where parking will be a problem, in which case light rail can get large crowds to and from the venue quickly and safely. Working with the police can help divert car trips to light rail. Finally, the light rail operator also can develop a social side, for example by securing sponsorship from a local beverage company to provide free or all-night service over the holidays, allowing people to go out with friends to enjoy themselves and get home safely.

8.7 Building a Network

A one-line light rail system can have a significant impact on reducing traffic congestion and the environmental impacts of traffic. Two lines however, with convenient interchange, will be more than twice as good, by virtue of network synergy. While the light rail promoter is getting the first line open, there will need to be some planning for subsequent lines (Figure 8.3), assuming that the first achieves the objectives set and is popular with the public. How this can be achieved was discussed in Chapter 3. In a city of about 250,000, a three-line network, with suitable park-and-ride, can make a significant impact on attracting transit riders. Depending on the topology of the area, there may be scope for a larger network, but there is also a danger of diseconomies of scale creeping in, with riders for the new line merely being abstracted from the existing light rail network.

Continual public monitoring is therefore an important aspect of light rail operation. In particular, identifying when a patronage plateau is reached indicates the danger of the zenith of the product life cycle. At this point, the operator will need to "relaunch" the system, with improvements that are more than cosmetic. This is when having a depreciation fund will pay off, since improvements can be achieved from the fare box. If another line is also part of the relaunch, then the operator is in a strong position to raise the funding needed for the new line. Light rail can be built for about a tenth of

Figure 8.3 Another new light rail vehicle ready to run (Photo: L. Lesley)

the cost of a metro or subway and achieve at least 90% of the same objectives. Indeed, light rail can provide more convenient access and faster and more competitive journeys for most of the trips people presently make by private car. With electricity from renewable power, light rail is also sustainable, nonpolluting, and helps to reduce dependence on imported oil and other fossil fuels.

Appendices

Appendix 1: Generalized Costs and the Value of Time as a Method of Patronage Forecasting

1.1 Introduction

Forecasting the patronage of a new (public) transport facility (e.g., new rail line, bus route, etc.) is important in the cost-benefit analysis often used to justify the investment. The accuracy of such forecasts has been criticized in earlier reports, and overly optimistic forecasts have led to reductions in public funding for new tramways and other public transport projects in the UK.

Part of the reason why forecasts are not accurate is that there are many empirical methods based on time series analyses. These rarely have proven cause-and-effect relationships, even though high statistical correlations prove the relationships to be nonrandom (Moroney 1981). Figure 1 shows a typical time series model, based on 20 years of data (1985 to 2005). By interpolation, the regression line through the data shows the midpoint (1995) to be the most accurately predicted value. Extrapolating 10 years beyond the latest data point (2005) shows the confidence interval widening, so that in 2015, while the forecast value is 11 million passengers annually, the probable range (at the 95% confidence level) lies between 8 and 18 million. Even at a lower (67%) confidence level, the forecast range is still 10 to 13 million passengers per year. This, however, cannot exclude autocorrelation, where two or more variables measure the same behavior but slightly differently.

Time series analysis cannot predict step changes in behavior, nor even prove that past behavior will continue into the future. To overcome this, two different approaches can be adopted. In the first, a combination of fuzzy logic

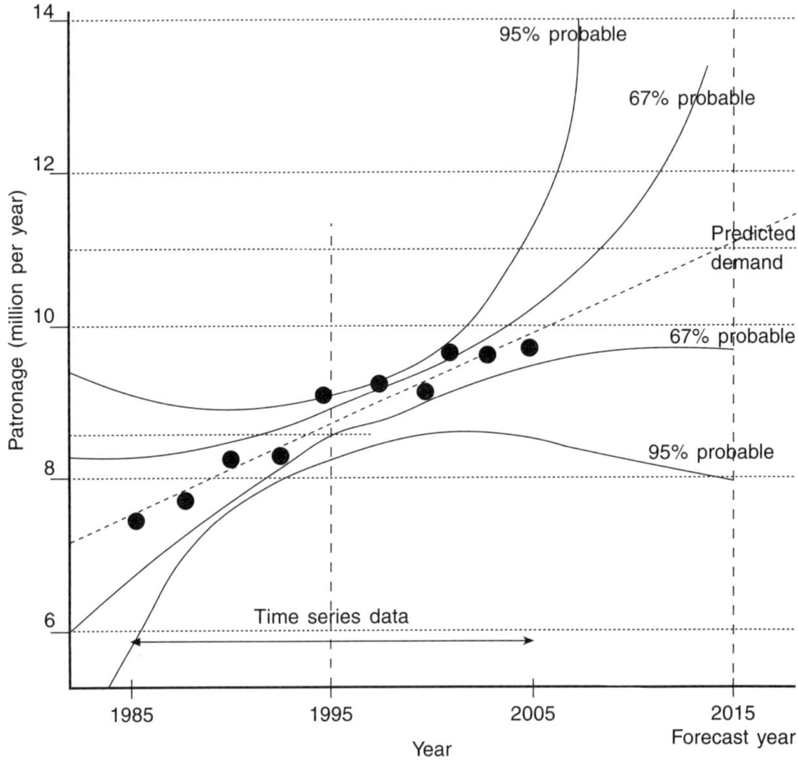

Figure 1 Time series analysis

and chaos theory can be used to identify possible behavior break points, which is the subject of a separate paper in preparation. In the second, a causal model based on individual behavior can be used to predict, independent of time series data, the modal split or patronage level. One method to achieve this is based on generalized costs, which models the perceived quality of different modes of transport.

1.2 Theory of Generalized Costs

It is well established in economic theory that there is an inverse relationship between demand and cost, the demand curve, shown in Figure 2 (Lipsey 1979) and represented mathematically as

$$N \propto \frac{1}{C}$$

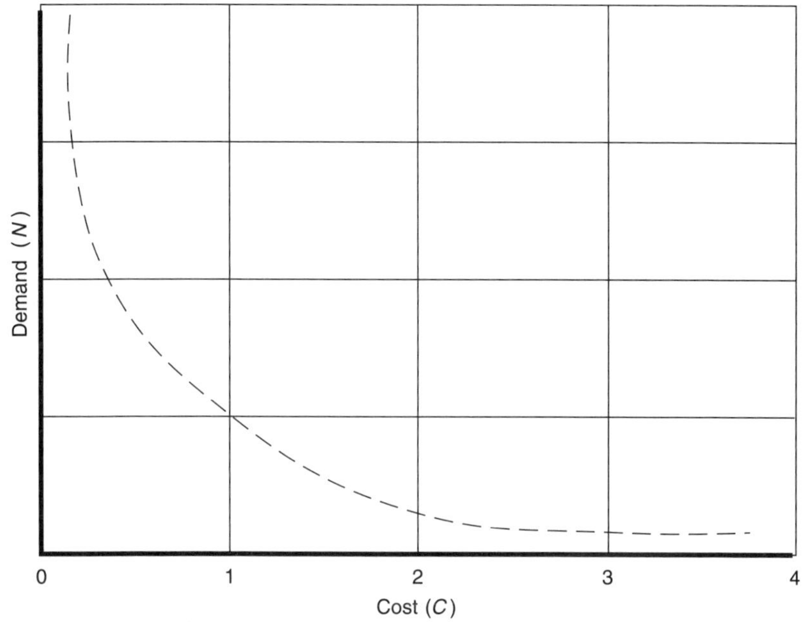

Figure 2 Economic demand (*N*)/cost (*C*) curve

where N = demand (e.g., number of trips) and C = cost (of trips). Therefore:

$$N = k \cdot \frac{1}{C}$$

where k = a constant.

Unfortunately, where the money cost of transport is substituted into this relationship, there is a weak correlation, and typically the elasticity of demand (e) is low:

$$e = \frac{dn}{dC} < 0$$

For urban bus systems in the UK, a figure for e has typically been –0.3. Recently, as bus use has declined further in the UK, e = –0.4 has been found from fare increases. These elasticity figures mean that between 60 and 70% of the change in patronage is determined by factors other than the money cost of travel (Mathisen 2006; Lesley 2004).

Through market research and behavioral studies, the other factors that influence mode choice in transport have been identified. These represent measures of the quality of service, speed of travel, ease of access, and reliability of service. How can these be measured, and are there relative weights to represent any bias in the perception by passengers (Transportation Research Board 1974; Stopher and Meyburg 1976)?

The journey time from origin to destination, as has been found from a large number of studies, can be separated into different parts. Each part has its own value and weighting. A typical urban public transport journey starts with a walk to the (bus) origin station and includes waiting for the departure, a ride on the vehicle, and then walking to the final destination. On some journeys, there may also be a change of route, requiring a walk between stops and a further wait for the new departure (Goodwin 1974; Hensher and McLeod 1974; Rogers et al. 1970).

Research shows that in-vehicle ride time is perceived by passengers at about the same as clock time. This means that a 10-minute ride in a bus feels like 10 minutes of clock time. Waiting time, however, has a much higher perceived value. Values as high as 2.2 × clock time have been reported. This means that every minute of waiting time saved (e.g., by more regular service) has the same benefit as reducing in-vehicle riding time (e.g., by faster running) by 2 minutes (Davies and Rogers 1973).

Walking (access) time is viewed more onerously than in-vehicle time but not as badly as waiting time. Values of 1.8 × clock time have been reported. This means that every minute of walking time saved (e.g., by having more stops) has the same benefit as reducing in-vehicle time by about 2 minutes. Passengers also value vehicle comfort and safety, but for most urban journeys their importance is rather small compared to the components of travel time.

Now we face a fundamental discontinuity. The components of travel time are measured in units of time, with international recognition, standardization, and interchangeability. The travel cost (of public transport) is, on the other hand, measured in units of money in the local currency, making comparisons with other countries difficult. Fortunately, further research shows that different reasons for traveling (journey purposes) have different values, which reflect the passenger's available income and the economic importance of the journey (Goodwin 1976; Bruzelius 1981).

This can be captured by a value of time, allowing the time elements of a journey to be converted into money or the (fare) costs to be converted into time. The most important travel for most passengers is the journey to work. Here research shows the perceived value of time to be close to the wage rate of the passenger. For other (nonwork) journeys, a variety of values have been reported, clustering around 25% of the wage rate. For convenience, an average of 25% can therefore be used for other (nonwork) journeys (Hyman and Wilson 1968; de Donnea 1972).

1.3 Establishing the Functional Relationships

To better understand these factors, two worked examples showing the components of an urban (public transport bus) work trip are presented.

Journey 1

Part of journey	*Time (minutes)*	*Weight*	*Equivalent minutes*
a. Origin walk	5	1.8	9.0
b. Wait at stop	5	2.2	11.0
c. Riding	12	1.0	12.0
d. Destination walk	6	1.8	12.8
		TOTAL	44.8

e. Fare $1.50 (at value of time $0.20/minute)

Journey 2

Part of journey	*Time (minutes)*	*Weight*	*Equivalent minutes*
a. Origin walk	4	1.8	7.2
b. Wait at stop	4	2.2	8.8
c. Riding	10	1.0	10.0
d. Destination walk	3	1.8	5.4
		TOTAL	31.4

e. Fare $1.50 (at value of time $0.20/minute)

1.3.1 Generalized Costs in Units of Money

Journey 1

a. Travel time = 44.8 minutes @ $0.20/minute =	$8.96
b. Fare =	$1.50
TOTAL	$10.46

Journey 2

a. Travel time = 31.4 minutes @ $0.20/minute =	$6.28
b. Fare =	$1.50
TOTAL	$7.78

1.3.2 Generalized Costs in Units of Time

Journey 1

a. Travel time =	44.8 equivalent minutes
b. Fare = $1.50 @ $0.20/minute	7.5 equivalent minutes
TOTAL	52.3 equivalent minutes

Journey 2

a. Travel time =	31.4 equivalent minutes
b. Fare = \$1.50 @ \$0.20/minute	7.5 equivalent minutes
TOTAL	38.9 equivalent minutes

In both methods of generalized costs determination, the money portion is small (about 15 to 25%), and as the quality of the journey worsens, the generalized costs increase.

If the modal choice analysis is related to only one period of time or only one country, then there is little choice between equivalent minutes or money as a measure of generalized costs. If, however, there is a historical perspective, forecasts are required, or there are international comparisons, then generalized costs should be measured in equivalent minutes.

1.3.3 Representing Generalized Costs

Generalized costs can now be represented as shown in Figure 3. In general terms:

$$N \propto \frac{1}{GC}$$

Substituting a constant of proportionality (k):

$$N = \frac{k}{GC}$$

If an initial condition of GC_1 and N_1 is known, the constant can be calculated. On the other hand, if a new operating condition GC_2 (e.g., a fare increase or improved quality of service) is known, then the new demand N_2 can be found by:

$$\frac{N_2}{N_1} = \frac{GC_1}{GC_2}$$

Therefore:

$$N_2 = N_1 \times \frac{GC_1}{GC_2}$$

This can now be applied to the examples set out above.

If GC_1 = 56.8 equivalent minutes and GC_2 = 43.4 equivalent minutes, then:

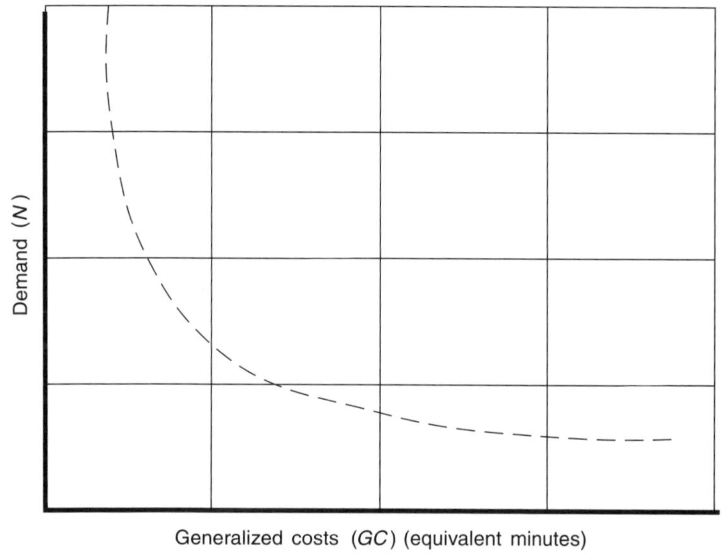

Figure 3 Generalized cost/demand curve

$$N_2 = N_1 \times \frac{56.8}{43.4}$$

Therefore:

$$N_2 = 1.308 \times N_1$$

The new demand will be about 30% higher than the original demand, when the generalized costs fall by 24%.

1.4 Introducing a New Mode of Transport

The above analysis is obviously valid where incremental changes are made to existing modes of transport (e.g., higher bus fares, faster car speeds, etc.). How can the introduction of a new mode of transport be handled with generalized costs, to provide patronage figures for the new mode and determine the reduction in use of existing modes (Strand 1999)?

The easiest way to consider this is by means of pair-wise probabilities. For small samples of data, there is a danger that probabilities will over- (or under-) estimate the new patronage. On the other hand, with large samples, or population sizes, the probability functions will be very close to the outcome average.

Let N_1 and GC_1 be the present demand for and generalized cost of existing mode 1, N_2 and GC_2 be for mode 2, etc. Now let GC_a be the generalized cost of a new planned mode of transport a. As

$$N_1 \propto \frac{1}{GC_1}$$

so

$$N_1 = k \cdot \frac{1}{GC_1}$$

and

$$N_1 + N_2 = k \cdot \frac{1}{GC_1} + k \cdot \frac{1}{GC} = k \cdot \frac{(GC_2 + GC_1)}{(GC_2 \times GC_2)}$$

To determine the probability of N_1 compared to N_2:

$$P_{1/2} = \frac{N_1}{(N_1 + N_2)} = k \cdot \frac{1}{GC_1} \bigg/ k \cdot \left[\frac{(GC_2 + GC_1)}{(GC_1 \times GC_2)} \right]$$

$$= \frac{GC_2}{(GC_1 + GC_2)}$$

Thus, pair-wise probabilities can be calculated, and the part of the demand attracted from mode 1 then becomes

$$P_{1/a} = \frac{GC_1}{(GC_1 + GC_a)}$$

If N_1 = the existing demand for mode 1, then the attraction from mode 1 to mode a will be

$$N_a = N_1 \times P_{1/a} = N_1 \cdot \frac{GC_1}{(GC_1 + GC_a)}$$

Similar pair-wise probability can be constructed for the attraction of demand to mode a from modes 2, 3, etc.

1.5 Application of Concrete Data

A corridor in a city is 10 km long and presently has the following modal split:

Mode	*Trips per year (million)*	*Average generalized cost (equivalent minutes)*
1. Bus	8.0	40
2. Car	9.0	35
3. Walk	2.0	70
TOTAL	19.0	

A new mode a of transport (e.g., tram) is planned for this corridor. The tram will have a lower generalized cost through more frequent/reliable service and faster operating speeds. Assume the new mode a has a generalized cost = 30 equivalent minutes.

Using the pair-wise probability analysis developed in Section 1.4, the attraction of existing trips to the new mode a can be calculated as follows:

$$N_{1/a} = 8.0 \times \frac{40}{(40 + 30)} = 4.6 \text{ million}$$

$$N_{2/a} = 9.0 \times \frac{35}{(35 + 30)} = 4.8 \text{ million}$$

$$N_{3/a} = 2.0 \times \frac{70}{(70 + 30)} = 1.4 \text{ million}$$

Total demand for new mode a = 10.8 million per year

Thus, the new tramway will attract about 50% of the present total corridor traffic. This ignores any traffic attracted from other (parallel) corridors or the generation of new trips (e.g., carrying disabled passengers in wheelchairs, etc.).

How sensitive are these values to the quality of the new (tram) mode? If the average generalized cost of the new mode was higher, at 35 equivalent minutes, how many trips would be attracted?

$$N_{1/a} = 8.0 \times \frac{40}{(40 + 35)} = 4.3 \text{ million}$$

$$N_{2/a} = 9.0 \times \frac{35}{(35 + 35)} = 4.5 \text{ million}$$

$$N_{3/a} = 2.0 \times \frac{70}{(70 + 35)} = 1.3 \text{ million}$$

Total demand for new mode a = 10.1 million per year

Increasing the new tramway generalized cost by 17% reduces the forecast tram traffic by 7%.

If, on the other hand, the tramway replaces the bus service, then either all existing bus passengers will transfer to the tram or there will be a redistribution of trips between car and walk and tram service.

Option 1: All bus passengers transfer to tram

From	*Demand per year (million)*
a. Bus	8.0
b. Car	4.8
c. Walk	1.2
TOTAL	14.0

Option 2: Redistribution of bus trips

a. Bus to car =

$$8.0 \times \frac{40}{(40 + 35)} = 4.3 \text{ million}$$

Therefore, new car demand =

$$9.0 + 4.3 = 13.3 \text{ million per year}$$

b. Bus to walk =

$$8.0 \times \frac{70}{(70 + 40)} = 5.1 \text{ million}$$

Therefore, new walk demand =

$$2.0 + 5.1 = 7.1 \text{ million per year}$$

c. Car to tram =

$$13.3 \times \frac{35}{(35 + 30)} = 7.2 \text{ million}$$

d. Walk to tram =

$$7.1 \times \frac{70}{(70 + 35)} = 4.7 \text{ million}$$

All bus traffic redistributed

New demand	*Volume per year (million)*	*Total (million per year)*
Tram	7.2 + 4.7	11.9
Car	13.3 – 7.2	6.1
Walk	7.1 – 4.7	2.4
Corridor TOTAL		20.4

The new tram demand is 15% less than assuming all bus trips transfer, and possibly more realistic. Alternatively, this can be considered as the low (11.9 million) and the other the high (14.0 million) forecast. More importantly, the tram will be carrying between 49 and 75% more passengers than the bus service it replaces.

1.6 A Practical Example from Galway, Ireland

Galway is a city of 85,000 on the west coast of Ireland. Presently 4% of all city trips are by bus, 79% by car, and 17% walk and cycle. There are only four bridges over the River Corrib, which splits the city into western (residential) and eastern (commercial and industrial) halves, with severe traffic congestion in the morning and evening peaks. At the feasibility study stage, seven route options were studied. From these, a two-line, 21-km-long tramway was planned to serve the most densely developed parts of the city. Sixty percent of residents live within 500 m of a new tram line. There will be a total of 50 stations, including 4 major interchange stations (Galway Group 2008).

The new tramway will operate in competition with two bus companies and a high level of car ownership (over 70% of households have a car). Generalized costs have been used to determine the likely patronage of a new mode (tram) of transport, to examine the sensitivity to data uncertainties and to examine how the tramway will cope under different competitive conditions.

Impact of reduced bus fares

	Bus fare = €2		*Bus fare = €1*	
Tram route options	*Tram passengers per year*	*Revenue per year (€ million)*	*Tram passengers per year*	*Revenue per year (€ million)*
1	801,000	2.4	780,000	2.3
2	1,412,000	4.2	1,403,000	4.2
3	763,000	2.3	741,000	2.2

Tram route options	*Bus fare = €2*		*Bus fare = €1*	
	Tram passengers per year	*Revenue per year (€ million)*	*Tram passengers per year*	*Revenue per year (€ million)*
4	227,000	0.7	218,000	0.7
5	708,000	2.1	682,000	2.1
6	960,000	2.9	914,000	2.7
7	545,000	1.6	527,000	1.6

Impact of changing car speeds

Tram route options	*Speed = 40 km/hour*		*Speed = 20 km/hour*	
	GLUAS passengers per year	*Revenue per year (€ million)*	*GLUAS passengers per year*	*Revenue per year (€ million)*
1	801,000	2.4	815,000	2.5
2	1,412,000	4.2	1,444,000	4.3
3	763,000	2.3	780,000	2.3
4	227,000	0.7	232,000	0.7
5	708,000	2.1	721,000	2.2
6	960,000	2.9	977,000	2.9
7	545,000	1.6	555,000	1.7

Impact of different tram operating speeds

Tram route options	*Slower = 20 km/hour*		*Base = 22 km/hour*		*Faster = 25 km/hour*	
	GLUAS passengers per year	*Revenue per year (€ million)*	*GLUAS passengers per year*	*Revenue per year (€ million)*	*GLUAS passengers per year*	*Revenue per year (€ million)*
1	733,000	2.9	735,000	2.9	738,000	3.0
2	1,286,000	5.1	1,293,000	5.2	1,303,000	5.2
3	685,000	2.7	690,000	2.8	695,000	2.8
4	203,000	0.8	205,000	0.8	207,000	0.8
5	645,000	2.6	645,000	2.6	647,000	2.6
6	877,000	3.5	877,000	3.5	878,000	3.5
7	499,000	2.0	498,000	2.0	497,000	2.0

Impact of different tram service frequency

	More trams = 5 minutes		Base = 6 minutes		Fewer trams = 10 minutes	
Tram route options	*Tram passengers per year*	*Revenue per year (€ million)*	*Tram passengers per year*	*Revenue per year (€ million)*	*Tram passengers per year*	*Revenue per year (€ million)*
1	741,000	3.0	735,000	2.9	703,000	2.8
2	1,300,000	5.2	1,293,000	5.2	1,231,000	4.9
3	693,000	2.8	690,000	2.8	652,000	2.6
4	206,000	0.8	205,000	0.8	194,000	0.8
5	652,000	2.6	645,000	2.6	616,000	2.5
6	887,000	3.6	877,000	3.5	838,000	3.4
7	504,000	2.0	498,000	2.0	477,000	1.9

Different combinations of the seven route options were considered, from which one network was selected as the most efficient for detailed planning.

Network profitability (internal rate of return)

Network	*Capital cost (€ million)*	*Operating cost (€ million)*	*Passengers per year*	*Revenue per year (€ million)*	*Internal rate of return*
A = 2 + 3, 6, 7	150	4.00	4,538,000	16.10	10.4%
B = 1 + 3, 5, 6	159	4.10	3,883,000	14.66	8.8%
C = 2 + 3, 5, 6	144	3.84	4,817,000	16.89	11.3%
D = 2 + 3, 4, 6, 7	185	4.77	3,180,000	10.05	5.0%

These analyses allowed the tram promoter to confirm the likely patronage and the business plan, so that the capital costs of €200 million were released.

1.7 Conclusion

Noncausal time series extrapolation models provide implausible methods for accurately forecasting the future patronage of public transport systems. This is particularly true when a new mode of transport (e.g., a tramway) is introduced (in competition with existing modes).

Generalized costs provide a causal model, linking the cost of using a (public transport) service and the demand created. Considerable research over the last 30 years has shown that the choice of a public transport mode includes measures of the quality, as well as the monetary costs of using the service. Generalized costs can be converted to standard units of measurement by a conversion factor: the value of time. This is sensitive to the purpose of the journey and the economic value gained by passengers. The journey to work is valued at about the hourly rate of pay of the passenger.

Generalized costs can be expressed in units of money or time (equivalent minutes). There are significant advantages in using a time-based measure, since this allows historic and international data to be compared directly. Generalized costs measured in equivalent minutes provide an inverse relationship with demand, directly equivalent to the classical economic demand curve. In addition to making patronage forecasts possible for changes to existing transport services, generalized costs also allow patronage forecasting when completely new modes of transport are introduced. This is possible using pair-wise probability analysis, comparing the generalized cost of an existing mode with that of the new mode. Generalized costs can be further adapted to examine the sensitivity of the accuracy of the input data and impact on the output forecasts.

Finally, generalized costs have been used over several decades for many real transport projects. A recent case study from Galway, Ireland showed how the promoters of the new tramway were able to evaluate the financial viability of a new tramway network. Crucial patronage and revenue forecasts were calculated from expected generalized cost values, by splitting the existing and known traffic demands in the corridors served.

Appendix 2: Households without and with Cars

		Households with car			
	No car	*Main driver*	*Other driver*	*Nondriver*	*All*
Trips per year					
Bus	184	13	45	78	39
Other transit	46	20	50	19	23
Taxi	25	5	8	9	7
All modes	775	1,229	1,013	918	1,096
Miles per year					
Bus	821	121	334	437	256
Other transit	774	654	1,081	365	607
Taxi	82	44	48	47	45
All modes	3,040	10,104	7,857	4,914	8,042

Source: UK National Travel Survey 2006, Department for Transport

Appendix 3: UK Fare Elasticities

Mode	*1980*	*2003*
Bus	–0.30	–0.4
Metro	–0.15	–0.3
Suburban rail	–0.50	–0.6

After Paulley et al. 2004

References

Anderson, N. (2009). Freight by tram—is there a case for it? *Tramways & Urban Transit,* Vol. 72, No. 854, February.

Anon. (1986). Cheaper track could make Alice-Darwin project viable, *Railway Gazette International,* Vol. 142, No. 3, p. 192.

Atkins, H.N. (1980). *High Materials, Soils and Concretes,* Reston Publishing.

Audit Commission (2006). Annual Audit Letter, November (merseytravel.gov.uk/pdf/MerseytravelAAL.pdf).

Bae, C.H.C., Jun, M.J., and Park, H. (2003). The impact of Seoul's subway Line 5 on residential property values, *Transport Policy,* Vol. 10, No. 2.

Banister, D. (2002). *Transport Planning,* Spon Press.

Bartlett, D. (2008). Cash waste over Merseytram after local infighting, *Liverpool Daily Post,* January 21.

Bonsall, P.W. (1997). Principles of transport analysis and forecasting, *Transport Planning and Traffic Engineering,* C.A. O'Flaherty, Ed., Arnold.

Bonsall, P.W. and O'Flaherty, C.A. (1997). Observational traffic surveys, *Transport Planning and Traffic Engineering,* C.A. O'Flaherty, Ed., Arnold.

Bruton, M.J. (1970). *Introduction to Transportation Planning,* Hutchinson.

Bruzelius, N.A. (1981). Microeconomic theory and generalized cost, *Transportation,* Vol. 10, No. 3.

Buchanan, C.D. (1963). *Traffic in Towns,* Report of Working Group, HMSO, London.

Buckley, R.J. (1975). *History of Tramways from Horse to Rapid Transit,* David & Charles.

Butler, E.W., Chapin, F.S., Hemmens, G., Kaiser, E.J., Steginan, M.A., and Weiss, S.F. (1969). *Moving Behavior and Residential Choice,* Report 81, Highway Research Board.

Bygate, J. (1995). South Yorkshire Supertram—first year of operation, Light Rail 95 Conference, Croydon, London, November 14–16.

Chase, C. (1998). Jolly trolley folly, *San Diego Earth Times* (www.sdearthtimes.com/cut_to_chase/ctc_on_trans.html), September 2.

Cheek, C. (2008). NTS data reinforces effects of car ownership on bus use, *Transit Magazine,* August.

Christaller, W. (1933). *Die zentralen Orte in Suddeutschland,* Gustav Fischer, Jena; *Central Places in Southern Germany,* translated (in part) by Charlisle W. Baskin, Prentice Hall, 1966.

Churchman, G.B. (1995). *Railway Electrification in Australia and New Zealand,* IPL Books.

Coffey, A. and Kuchwalek, H. (1992). Grunes licht fur die strassenbahn, verlag fur Gesellschaftskritik,.

Cudahy, B.J. (1995). *Cash, Tokens and Transfers,* Fordham University Press.

Cuffe, P. (1989). Control and communications, Light Rail 89, International Conference, Liverpool Polytechnic, November.

Daly, A. (1981). Behavioural travel modelling, *Transport & Public Policy Planning,* Mansell, London.

Davies, A.L. and Rogers, K.G. (1973). *Modal Choice and the Value of Time,* Local Government Operation Research Unit Report C.143.

de Donnea, F. (1972a). *The Determinants of Transport Mode Choice in Dutch Cities,* Rotterdam University Press.

de Donnea, F.X. (1972b). Consumer behaviour, transport mode choice and value of time: some micro-economic models, *Regional and Urban Economics,* Vol. 1, No. 4.

Dewees, D.N. (1976). The effect of a subway on residential property values in Toronto, *Journal of Urban Economics,* Vol. 3, No. 4.

Dix, M. (1981). Contributions of psychology to developments in travel demand modelling, *Transport & Public Policy Planning,* Mansell, London.

Dunbar, C.S. (1967). *Buses, Trolleys & Trams,* Paul Hamlyn.

Faludi, A.K.F. (2004). Impact of a planning philosophy, *Planning Theory,* Vol. 3, No. 3, pp. 225–236.

Galway Group (2008). www.GLUAS.com.

Garbutt, P. (1989). *World Metro Systems,* Capital Transport.

Glaister, S. (1981). *Fundamentals of Transport Economics,* St. Martin's Press, New York.

Golob, T.F., Canty, E.T., and Gustafson, R.L. (1972). *Transportation Research,* Vol. 6, No. 1.

Goodwin, P.B. (1976). *Value of Time,* Report of 30th Roundtable on Transport Economics, European Conference of Ministers of Transport, Paris.

Goodwin, P.B. (1993). Car ownership and public transport use: revisiting the interaction, *Transportation,* Vol. 20, No. 1.

Goodwin, P.H. (1974). Generalized cost, time and the problem of equity in transport studies, *Transportation,* Vol. 3, No. 1.

Hass-Klau, C. (2002). *Future of Urban Transport,* ETP.

Hass-Klau, C. (2003). *Bus or Light Rail: Making the Right Choice,* 2nd edition, ETP.

Hass-Klau, C. (2004). *Economic Impact of Light Rail: The Results of 15 Urban Areas in France, Germany, UK and North America,* ETP.

Hay, A. (1986a). *The Impact of Subsidised Low Fare Public Transport on Travel Behaviour,* TRRL RR38.

Hay, A. (1986b). The impact of subsidised low-fare public transport on travel behaviour, *Environment and Planning C: Government and Policy,* Vol. 4, No. 2, pp. 233–246.

Hensher, D.A. (1974). *Consumer Preferences in Urban Trip-making,* Commonwealth Bureau of Roads, Australia.

Hensher, D.A. (1989). Behavioural and resource values of travel time savings, *Australian Road Research,* Vol. 19, No. 3.

Hensher, D.A. and McLeod, P.B. (1974). *Generalizing and Generalized Cost Functions,* Melbourne, Australia, December.

Hills, P.J. (1981). Does urban bus transport have a future? *Transportation,* Vol. 10, No. 4.

Hondius, H. (2007). Combine rectification programme moves ahead, *Railway Gazette International,* May 1.

Horne, J.B. and Maund, T.B. (1982). *Liverpool Transport,* Vol. 2: 1900–1930, Transport Publishing.

Horvath, A., Kerkapoly, E., and Megyeri, J. (1977). *Kulonleges Vasutak,* Muszaki Konyvkiado, Budapest.

Hovell, P.J., Jones, W.H., and Moran, A.J. (1975). *The Management of Urban Public Transport,* Saxon House.

Hubbell, J. and Colquhoun, D. (2006). Light rail transport in Calgary, the first 25 years, presented at the TRB International Light Rail Conference, St. Louis, April 8–12.

Hupkes, G. (1982). The law of constant travel time and trip rates, *Futures*, Vol. 14, No. 42.

Hyman, G.M. and Wilson, A.G. (1968). *The Effects of Changes in Travel Costs on Trip Distribution and Modal Split*, Working Paper 21, Centre for Environmental Studies, December.

Jones, B. (2007). South Yorkshire's quiet success story, *Tramways & Urban Transit*, No. 834, June.

Joyce, J. (1964). *Modern Tramway Review*, Ian Allen.

Keenan, D. (1979). *Tramways of Sydney*, Transit Press.

Laconte, P. (1978). Marketing and Public Transport, UITP Conference, Newcastle upon Tyne.

Langdon, M.G. (1971). *Cabtrack Studies*, Royal Aircraft Establishment, Ministry of Defense.

Latour, B. (1971). The Cabtrack urban transport system, *Traffic Engineering & Control*, Vol. 2, pp. 634–638.

Lesley, L. (1974a). Improving the image, *Motor Transport, Bus & Coach Supplement*, March 1.

Lesley, L. (1974b). *The Image of Public Transport in Two Housing Estates in the Teesside Conurbation*, Action Report No. 4, University of Newcastle upon Tyne, August.

Lesley, L. (1974c). *The Effects of Advertising Bus Services in Two Housing Estates in the Teesside Conurbation*, Action Report No. 5, University of Newcastle upon Tyne, August.

Lesley, L. (1976). Optimum bus stop spacing, *Traffic Engineering & Control*, Vol. 10 (part 1), Vol. 11 (part 2).

Lesley, L. (1983). Runcorn—a rapid transit new town, *Built Environment*, Vol. 9, No. 3/4.

Lesley, L. (1986). Public transport in Budapest—some observations about future developments, *Periodica Polytechnica*, Vol. 14, No. 1.

Lesley, L. (1987). Budget light rail—a product of the discount rate, *Electrifying Urban Public Transport*, Conference Proceedings, Liverpool Polytechnic, November.

Lesley, L. (1991). *Light Rail Handbook*, Transport Science Ltd.

Lesley, L. (1996). Public & private funding in new urban public transport, *Transport in Transition*, Public Finance Foundation.

Lesley, L. (2002). Testing the next generation of tramway tracks, Railway Engineering 2002 Conference, London, July 3.

Lesley, L. (2004). Forecasting traffic demand, Guest lecture, National University of Ireland, Maynooth, March 24.

Lesley, L. (2005). *Affordable Mass Transit? Mass Transit*, Washington D.C., February.

Lesley, L. (2006). New OHL system, *The Rail Engineer*, No. 24, October.

Lesley, L. (2009a). Fatigue in railway and tramway track, *Fatigue in Railway Infrastructure*, Woodhead Publishing.

Lesley, L. (2009b). Generalised costs and the value of time as a method of patronage forecasting, *Acta Technica Jaurinensis*, Vol. 2, No. 1.

Lipsey, R.G. (1979). *An Introduction to Positive Economics*, Weidenfeld & Nicolson, London, chap. 7.

Lowson, M.V. (1999). Cabtrack, *Proceedings Institution of Civil Engineers Transport*, Vol. 135, No. 3, pp. 139–151.

Mathisen, T.A. (2006). *The Relationship between Travel Distance and Fares, Time Costs and Generalized Costs in Passenger Transport*, NO-8049, Bodø Graduate School of Business, Norway.

Matsoukis, E.C. (1980). Speed/flow relationships in the central Glasgow area, *Transportation Planning and Technology*, Vol. 6, No. 1, pp. 21–26.

McLuhan, M. (1967). *The Medium Is the Message*, University of Toronto Press.

McMillan, R.K. and Assael, H. (1968/9). *National Survey of Transportation Attitudes and Behavior*, Report No. 49 and 82, Highway Research Board, Washington, D.C.

Mogridge, M.J.H. (1990). *Travel in Towns: Jam Yesterday, Jam Today and Jam Tomorrow?* Macmillan Press, London.

Morgan, B. (1973). *Railways: Civil Engineering*, Arrow.

Moroney, M.J. (1981). *Facts from Figures*, Penguin Science Books, London.

National Audit Office (2004). *Improving Public Transport in England through Light Rail*, England.

O'Flaherty, C.A. (1997). *Transport Planning and Traffic Engineering*, Arnold.

Paine, F.T., Nash, A.N., Hille, S.J., and Brunner, G.A. (1969). Attributes in bus transportation: importance depends on trip purpose, *Journal of Applied Psychology*, Vol. 53, No. 6.

Pas, E.I. and Harvey, A.S. (1997). Time use research & travel analysis modelling, *Understanding Travel Behaviour in an Era of Change*, Pergamon.

Paulley, N., Mackett, R., Preston, J., Wardman, M., Titheridge, H., and White, P. (2004). Factors affecting the demand for public transport, European Transport Conference.

Pharoah, T. and Apel, D. (1995). *Transport Concepts in European Cities*, Ashgate Publishing.

Pitrel, J. (1987). Nantes modern tramway—first results and forecasts, Proceedings Electrifying Urban Public Transport Conference, Liverpool, November 17–18.

Powell, M.F. (1997). *Light Rail & Guided Transit,* Natula Publications.

Profillidis, V.A. (2000). *Railway Engineering,* Ashgate.

Rogers, K.G., Townsend, G.M., and Metcalf, A.E. (1970). *A Generalized Cost Explanation of a Modal Choice Planning for the Journey to Work,* Local Government Operation Research Unit Report C.67, April.

Salkield, T. (1953). *Road Making and Road Using,* Pitman, London.

Sharp, D.R. (1970). *Concrete in Highway Engineering,* Pergamon Press.

Skamris, M. and Flyvbjerg, B. (1997). Inaccuracy of traffic forecasts and cost estimates on large transport projects, *Transportation Policy,* Vol. 4, No. 3.

Slinn, M., Matthews, P., and Guest, P. (1998). *Traffic Engineering Design,* Arnold.

Snowdon, J. (2007). Croydon: track renewal breaks the mould, *Tramway and Urban Transit,* Vol. 70, No. 836.

Spear, B.D. (1976). Attitudinal modeling: its role in travel-demand forecasting, *Behavioral Travel-Demand Models,* Transportation Research Record 592, Washington, D.C.

Stanley, J.K. (1977). Behavioral modeling: an evaluative perspective, 3rd International Conference on Behavioral Modeling, South Australia.

Starkie, D.N.M. (1976). *Transportation Planning, Policy and Analysis,* Pergamon Press.

Stopher, P.R. and Meyburg, A.H. (1976). *Behavioral Travel-Demand Models,* Lexington Books.

Strand, S. (1999a). *Bottlenecks in Transport Research,* Report 431/1999, Institute of Transport Economics, Oslo.

Strand, S. (1999b). Airport-specific traffic forecasts: the resultant of local and non-local forces, *Journal of Transport Geography,* Vol. 7, No. 1.

Streeter, T. (2009). Dresden: innovation is born from experience, *Tramways & Urban Transit,* No. 857, May.

Tennyson, E.L. (1989). *Impact on Transit Patronage of Cessation or Inauguration of Rail Service,* Transportation Research Record 1221, Washington, D.C.

Tolley, R. and Turton, B. (1995). *Transport Systems, Policy and Planning,* Longman.

Townsend, E.C. (1969). *Investment and Uncertainty,* Oliver & Boyd.

Transportation Research Board (1974). *Behavioral Demand Modeling and the Valuation of Travel Time,* Report No. 149, Proceedings of C Conference, South Berwick, Maine, July 1973.

Transportation Research Board (1992). *Light Rail Transit: Planning, Design, and Operating Experience,* Transportation Research Record 1361, papers presented at the Sixth National Conference on Light Rail Transit, Calgary, Alberta, May 24–27.

Transportation Research Board (2009). *Track Design Handbook for Light Rail Transit*, TCRP D-14, Washington, D.C., March.

US Department of Transportation (1976). *Effective Citizen Participation in Transportation Planning*, 2 vols., US Government Printing Office, Washington, D.C.

Vickerman, R.W. (1991). *Infrastructure and Regional Development*, Pion Press.

Voorhees, A.M. et al. (1974). *Public Transport in Bedfordshire: Facts, Views, Issues, Options*, Alan Voorhees & Associates.

Wachs, M. (1985). The politicization of transit subsidy policy in America, *Transportation and Mobility in an Era of Transition*, Elsevier Press.

White, P. (2005). *Public Transport, Planning, Management & Operation*, UCL Press.

www.apta.com/sites/transus/lightrail.htm.

www.lightrail.com.

www.lightrail.net.

www.lightrailnow.org.

www.LRTA.org.

www.transport2000.ca.

www.vtpi.org.

www.wikipedia.org/wiki/Light_rail.

Index